Audi-Dissertationsreihe, Band 116

A multifactorial analysis of thermal management concepts for high-voltage battery systems

Von der Fakultät für Maschinenbau
der Technischen Universität Carolo-Wilhelmina zu Braunschweig

zur Erlangung der Würde eines
Doktor - Ingenieurs (Dr. - Ing.)

genehmigte Dissertation

von: Joshua Smith
aus (Geburtsort): Vail, USA

Referenten: Prof. Dr. - Ing. Jürgen Köhler
                   Prof. Dr. - Ing. Christoph Herrmann
Vorsitzender: Prof. Dr. - Ing. Ferit Küçükay

Eingereicht am: 13. November 2015
Mündliche Prüfung am: 02. Juni 2016

2016

Bibliografische Information der Deutschen Nationalbibliothek
Die Deutsche Nationalbibliothek verzeichnet diese Publikation in der
Deutschen Nationalbibliografie; detaillierte bibliografische Daten sind im Internet
über http://dnb.d-nb.de abrufbar.
1. Aufl. - Göttingen: Cuvillier, 2016
    Zugl.: (TU) Braunschweig, Univ., Diss., 2016

© CUVILLIER VERLAG, Göttingen 2016
    Nonnenstieg 8, 37075 Göttingen
    Telefon: 0551-54724-0
    Telefax: 0551-54724-21
    www.cuvillier.de

    ISBN 978-3-7369-9359-4
  eISBN 978-3-7369-8359-5

Acknowledgements

This research was performed during my time as an employee at AUDI AG, with additional financial support from the German Federal Ministry of Education and Research (BMBF) within the context of the research project *eProduction*. The academic mentoring from the Institute for Thermodynamics (IfT) at the University of Braunschweig combined with the support of my colleagues at Audi, as well as the project partners from *eProduction* all proved invaluable to the success of this work.

Thank you to Professor Dr.-Ing. Juergen Koehler from the Institute for Thermodynamics (IfT) at the University of Braunschweig for the academic mentoring and constructive discussions which helped this work take shape. Thank you to Professor Dr.-Ing. Christoph Herrmann from the Institute of Machine Tools and Production Technology (IWF) at the University of Braunschweig for acting as a reviewer of this Dissertation. Thank you to the review committee for their engagement.

Thank you to the e-production project partners in our work group for their support in the production of the technology demonstrators: Andreas Roth, Marc Essers, Uwe Maurieschat, Christian Schuch, Randeep Singh, Klaus Hermann, Jörg Damaske, and Andreas Track.

I am thankful for the support of TLK-Thermo GmbH, especially Dr.-Ing. Wilhelm Tegethoff and Dr.-Ing. Nicholas Lemke for the countless discussions, in person and over the phone. Their input was invaluable for determining the direction of the research.

Thank you also to my PhD-candidate-colleagues Bernhard Rieger, Jan-Christoph Menken, Thomas Weustenfeld, Dennis Ribbe-Sorano, and especially Pamina Bohn for the numerous discussions and continuous support.

My sincere thanks go to all my Audi colleagues who made this research possible, especially the work of masters students Peter Hable and Christoph Schneider, the experimental support of Uwe Opitz and Giacomo Gagliano, and the technical support of the production team. I will forever be thankful to Michael Hinterberger for his exceptional support and mentoring through all of our challenges, and to Christian Allmann for ensuring I had the time and resources to realize my goals.

Thank you also to the numerous colleagues and friends that also helped me achieve this goal, but that I have not named here.

Finally thank you to my family and fiancé Julia for their support throughout my studies and the process of completing this research.

Ingolstadt, September 2015

Joshua Smith

Contents

—————— Methodology ——————

—————— Analysis ——————

——— Appendices ———

1. Introduction

Governments, in both developed and developing nations, have acted to improve vehicle emissions standards. Health concerns in the 1970s first resulted in limitations on volatile organic compound, carbon monoxide and nitrogen oxide emissions from motor vehicles [c2es 2014, EPA 1999]. In recent years, the focus of regulation has shifted to carbon dioxide (CO_2) because of its effect on global climate change. Various nations have enacted taxes on large engines and provided tax incentives for purchasing alternative fuel vehicles. Legislation limiting the fuel consumption and CO_2 emissions of motorized vehicles has tightened; however, the newest generation of regulations driven by the United States, the European Union, and China present not only tougher fuel consumption and CO_2 reduction standards, but also formidable financial penalties if manufacturers fail to comply. This wave of legislation spurred by the fight against climate change has arguably instigated the recent surge in development of electric and hybrid-electric vehicles.

Directly comparing the legislation in various countries is difficult because of the various standard types (e.g. fuel economy, CO_2 standard), test cycles (e.g. NEDC, US-combined) and diverse calculation methods (e.g. vehicle weight, footprint) implemented. Additionally, the various credits and exemptions available leave room for interpretation and lobbying [Arena 2014, KPMG 2010]. The calculation method of the fuel economy of electric and hybrid-electric vehicles batteries also varies between nations and throughout time (e.g. during the phase-in of the requirements) [NHTSA 2011]. Regulations also vary based on vehicle size (e.g. vans, heavy duty), but only the regulations effecting passenger cars are discussed here. Regulations for key markets are summarized in Table 1.1.

Table 1.1: Comparison of upcoming fuel consumption / CO_2 regulations including the target year, standard type, the fleet target, calculation method, test cycle and penalties [Arena 2014].

Country / Region	Target Year	Standard Type	Unadjusted Fleet Target[1]	Calculation Method	Penalties
European Union	2015	CO_2	130 gCO_2/km	Weight based corporate average	Fines
	2020		95 gCO_2/km		
United States	2016	Fuel Economy and $GHGs^2$	36.2 mpg or 225 gCO_2/mi	Footprint-based corporate average	Fines
	2025		56.2 mpg or 225 gCO_2/mi		Sales Restriction
China	2015	Fuel Consumption	6.9 L/100 km	Weight based corporate average	Fines
	2020		5 L/100 km		Public proclamation

1) Set considering test cycle used
2) Greenhouse gas

As an example of the potential severity of the fines, manufacturers failing to meet the fleet average of 95 g CO_2 per kilometer in 2021 in the European Union will face fines of 95€ per gram over the 95 g CO_2 /km target for each registered vehicle. In China, import and sales bans can be hung on manufacturers. In the United States, the Environmental Protection Agency has the authority to access penalties up to $37,500 per vehicle [EPA 2009].

Though upcoming regulations are the most stringent yet, clauses exist to alleviate some of the pressure on manufacturers. In the United States and the European Union, small and midsize manufacturers are not subject to the full scope of the rules. Combinations of manufactures can also form "pools" amongst each other to reduce their fleet emissions. Innovations that effect the overall efficiency of the vehicle but are not directly measured through the test cycle can be credited toward the fleet average in the United States (Off-cycle credits) and the European Union (Eco-Innovations). These flexibilities are summarized in Table 1.2.

Table 1.2: Comparison of exemptions and flexibilities in major markets [Arena 2014].

	USA	EU	China
Derogation for small or middle-volume manufacturer	√	√	
Pooling[1]	√	√	√
Advantages for flexible fuel[2] and alternative fuel[3] vehicles	√	√	√
Advantages for advanced technology vehicles[4]	√	√	√
Eco-innovations[5]	√	√	
Banking and trading CO_2 emissions credits	√		

1) *Manufacturers may form a pool for the purpose of meeting fuel and/or emissions targets*
2) *Run on both conventional and alternative fuel*
3) *Run on exclusively alternative fuel*
4) *Electric vehicles, fuel cell vehicles and the plug-in portion of hybrid electric vehicles*
5) *Innovative technologies not captured in the test cycle*

Until now, efficiency improvements to the internal combustion engine and various vehicle components have allowed manufactures to meet the emissions standards. However, even with the efficiency improvements in gasoline and diesel engines, as well as the introduction of bio-fuels (E5, E10, E85) and natural gas (CNG/LNG), further alternatives are required to meet upcoming targets, as shown in Figure 1.1.

The so-called hybrid electric vehicle (HEV) has been applied throughout most vehicle segments to reduce fuel consumption, either with pure electric range or to support the internal combustion engine (ICE). Further electrification of the vehicle to decrease fuel consumption, such a 48 V on-board electronics or so-called mild-hybridization can also be implemented with conventional ICEs. Recently, plug-in hybrid (PHEV) technology has grown in significance because of the greater fuel saving potential, especially over the emissions test cycles currently used. PHEVs can utilize grid energy via an external plug, and have a more significant (10-50+ km) electric range. Combined with a gasoline or diesel engine, HEV and PHEV vehicles allow for fuel savings at a reduced cost compared to a fully-electric vehicle (BEV). BEVs can theoretically produce no emissions when charged with electricity generated from renewable energy, but this technology has been slow to gather a significant market share, most likely due to the high costs of the battery systems. Hydrogen fuel cell systems have been touted as the long–term solution to increased range, but they still rely on at least a small battery system [Sterner 2014a].

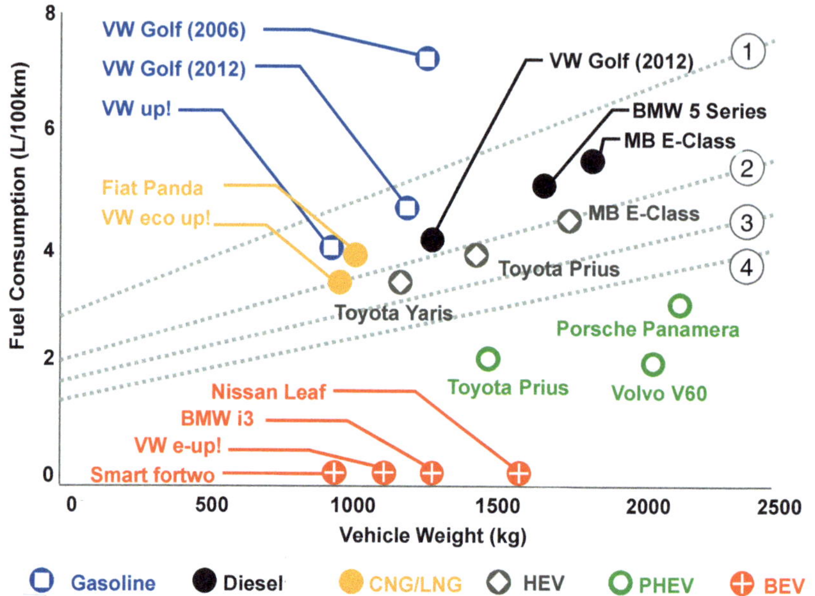

Figure 1.1: Various vehicle models grouped by drivetrain type plotted based on their fuel consumption (y-axis) and weight (x-axis) in relation to the European Union (EU) fleet targets (diagonal lines) in the years 2015 (1), 2020 (2) and proposals for 2025 (3,4). In the case of alternative fuels, a gasoline-equivalent consumption is calculated based on the energy content of one liter of gasoline [ICCT 2014].

Figure 1.1 shows the impact of PHEV and BEV vehicles in meeting EU fleet targets. Considering the development cycles for new vehicles and the recent surge in the electrification (mild-hybrid, HEV, PHEV and BEV), these advanced technology credits (see Table 1.2) will play the largest role in helping manufacturers meet the regulatory fleet targets set for the year 2020. Thereafter, as the credits are phased out, a larger fleet of fuel efficient vehicles will be required to meet regulations beyond 2020 world-wide.

1.1. Introduction to Lithium-Ion Battery Cells

The critical component in current electrified vehicles is the battery cell. Currently, Lithium-Ion cells are used. This technology was developed and successfully commercialized in consumer electronics; however, these so-called "consumer cells" were designed for small, portable devices with short lifetimes, requiring low power and designed for operation in mostly mild environments. Vehicles, on the other hand, must operate in various ambient conditions over surfaces of differing quality. Temperature extremes and rough roadways exert thermal stresses and mechanical vibrations for which most consumer electronics were never designed. Automotive cells are also orders of magnitude larger and have much higher capacities (up to 50 Ah). The target lifetimes of consumer cells (two to five years) are much less than the eight to ten-plus years required for the automotive industry [Lamp 2013].

In the quest for a automotive specific cell, various cell shapes, sizes and designs have been developed. Although the applied cell chemistries are by no means identical, they share general features. In contrast to galvanic cells (e.g. lead-acid battery) which contains solid and aqueous material states, a Lithium-Ion battery utilizes ordered structures at the positive and negative electrodes between which Lithium Ions move (intercalation). The positive electrode generally consists of a Lithium compound bound to aluminum foil, while the negative electrode consists of a carbon or graphite-based material bound to copper foil. The electrodes are separated by an electrically insulating polymer that allows the Ion transport (e.g. polypropylene or polyethylene) and surrounded by an electrolyte that aids the Ion transport. As secondary batteries, charging and discharging are repeatedly possible.

The active material used at the positive electrode and the carbon/graphite material used at the negative electrode help determine the electrochemical potential and correspondingly the energy density of the cell. Popular active materials include Lithium Cobalt Oxide ($LiCoO_2$), Lithium Ion Phosphate (LFP), Lithium Manganese Oxide (LMO), and Lithium-Nickel-Manganese-Cobalt Oxide (NMC). Different cell designs have different potentials and usable capacity ranging from 2.3-3.9 V and 110-190 Ah^1kg^{-1}, respectively [Sterner 2014b]. Additionally, different cell types respond differently to temperature and other external influences. The current standard in automotive applications is the NMC cell, as the combination of power density, lifetime, and the safety is best [Lamp 2013]. Regardless of the exact cell chemistry, Lithium-Ion cells are susceptible to various forms of aging, which results in a reduced vehicle range and battery system lifetime [Liaw 2013, Rao 2011, Vetter 2005]. Cell aging consists generally of two components: *calendrical* and *cyclical* aging [Hofmann 2014]. Calendrical aging, or aging over time, is a function of cell state of charge (SOC) and temperature. Cyclical aging, or the aging occurring during cell use, is a function of the charge / discharge rate, SOC, depth of discharge (DOD) and temperature [Sterner 2014b, Bandhauer 2014]. The focus of this research is on temperature-induced aging.

Cell aging and performance are influenced by temperature. The internal cell resistance increases with decreasing ambient temperature, especially near freezing and below. At such low temperatures (below 0°C), the cell suffers the most damage during charging due to so-called Lithium plating, which can also lead to safety risks in extreme cases [Legrand 2013, Zeyen 2013]. At such low temperatures, an inefficient discharge process also makes operation of the vehicle inefficient; however, discharging is less damaging than charging in this temperature range.

At high temperatures, the kinetics of the cell are actually accelerated, with the resulting drop in internal resistance causing an increase in available power; however, changes in the interplay between electrodes and electrolyte as well as the anode or cathode and electrolyte cause accelerated aging [Vetter 2005]. The rate of aging at temperatures above the optimal range can be approximated by the Arrhenius equation: for every 10 K temperature increase, the cell lifetime is halved [Zeyen 2013]. As the cell temperature climbs toward 90°C, electrolyte decomposition can occur, and as the temperature approaches 120°C, loss of the chemical reaction (so-called thermal runaway) is possible [Bandhauer 2011].

Not only is the average cell temperature critical, but also the temperature gradient over the cell and between cells. Uneven heat distribution has been shown to contribute to premature aging and capacity degradation by creating local "hotspots" [Lamp 2013, Fleckenstein 2012, Pesaran 2013, Kim 2007]. Temperature differences between cells cause the cells to age at different rates due to the electrical circuitry: in a chain of cells connected in series, the weakest cell lowers the system voltage [Koehler 2013]. For this reason, both the average temperature and the temperature differences must be considered in battery thermal management system (BTMS) design. The temperature dependency is explained in greater detail in Section 2.1.

Furthermore, with three different cell layouts (cylindrical, prismatic and pouch) available in a plethora of cell geometries and capacities, a multitude of unique thermal management challenges have been created. The internal layout of the cell influences the heat generation and therefore aging [Bandhauer 2011]. The various shapes require different solutions for contacting the thermal management system to the cell. Workgroups are in place to determine a single international standard (SO/IEC PAS 16898) for electric vehicles because currently various standards exist, including the 18650 cylindrical cell and the PHEV-2 prismatic cell (DIN Standard 91252) considered in this research [Lamp 2013].

The cell alone is by no means suited for vehicle integration: to function, automotive Lithium-Ion cells required a monitoring and control system, a mechanical housing, and connections to other high-voltage components [Sterner 2014b, Hofmann 2014, Koehler 2013]. Depending on the application, a thermal management strategy may also be required. Most vehicle manufacturers therefore rely on an intermediary building block, a so-called battery module, shown in Figure 1.2 [Koehler 2013, Kampker 2014].

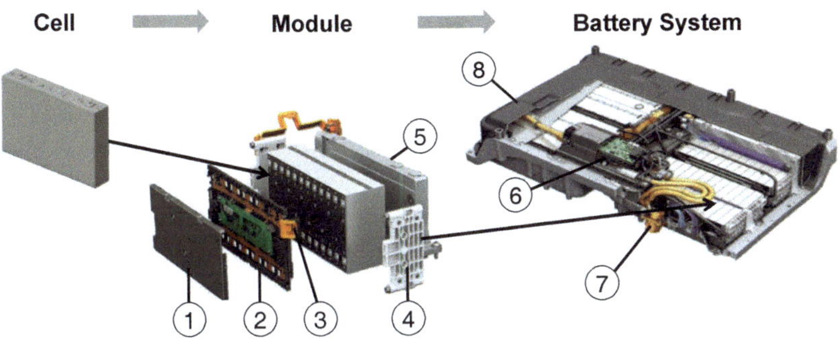

Figure 1.2: Levels of abstraction of a typical battery system: a cell (left) integrated in to a module (center) showing the cell module controller and integrated electrical contacts (1+2), the electrical connection between modules (3), the mechanical compression (4) and a cooling plate (5). The battery system (right) consisting of multiple modules housed in a casing (8) with additional power electronics (7) and the high-voltage connection to the vehicle (6). Parts adopted from [Audi 2014].

A battery module consists of any number of cells. The number of cells, combined with the circuitry chosen to connect them, determines the current and voltage of the module. Due to more stringent worker safety requirements above 60 V, the production process may be simplified by keeping the voltage below this threshold as long as possible in the assembly of a battery system. An additional consideration may be a module layout that can be used for future 48 V on-board circuitry. Regardless of the module size chosen, the same basic requirements must be fulfilled to facilitate safe operation. First, the cells must be contacted over their terminals electrically; however, no other portion of the cell may come in direct contact with another cell due to the potential for short circuiting. Additionally, some cells types exert high mechanical forces, requiring a compressive member around the cells. Depending on the location of the battery system in the vehicle, the casing required may need to absorb high crash loads, dampen vibration, and protect against corrosion from road debris [Lamp 2013, Koehler 2013]. Furthermore, a monitoring and control system is especially important for Lithium-Ion cells, which are prone to over-charging, deep discharging and short circuiting [Hofmann 2014]. Access for service or maintenance, recycling, and other end-of-life procedures must also be considered in battery system design. Depending on the vehicle use case, a thermal management strategy may be necessary, requiring the integration of additional components within the module, battery system, and vehicle. The integration of Lithium-Ion cells in a vehicle is a complex process, requiring expertise in various fields of technology.

The goal of battery thermal management is to protect against undesirable temperatures and temperature gradients across battery systems and individual cells in order to extend system lifetime, increase vehicle range, and protect against dangerous thermal states. However, thermal management systems can add size and weight to a battery system or require increased energy consumption (e.g. a fan or coolant pump), resulting in a reduction of vehicle range. Including a thermal management system generally adds cost and production complexity to the battery system as well. A multifactorial analysis considering for example the thermal performance, energy consumption, vehicle suitability, production complexity, and cost of various thermal management systems would allow for the identification of a suitable solution for a particular application. Additionally, given the rapid development of electric and hybrid-electric vehicles necessary to meet upcoming fuel economy and CO_2 regulations, an efficient method is required to analyze new and existing thermal management options as new cell types and vehicles are developed.

This research presents a novel solution to this challenge, including a method of analysis followed by a multifactorial analysis of various existing and new thermal management concepts. Existing experimental methodologies (state of the art) are presented in Section 3, while the state of the art in battery thermal management concepts as well as the selection of concepts to be analyzed within the scope of this research is presented in the following section.

1.2. State of the Art of Battery Thermal Management

Based on patents, publications, and experience in the field of battery thermal management, a brief overview of the state-of-the-art for the thermal management of Lithium-Ion automotive cells is discussed. Thereafter, the selection and classification of the concepts analyzed in this research are presented. Broadly speaking, thermal management can be characterized as either passive or active.

State of the Art of Passive Thermal Management

Passive thermal management is defined as utilizing no additional heating or cooling source besides the ambient [Buford 2011]. This includes the absence of a thermal management system. Passive thermal management strategies include insulating the battery pack from nearby heat sources or sinks, which has been proven to sink thermal management needs significantly, especially in adverse climates [Jayaraman 2011]. Additionally, connections with the ambient can be realized over a coolant loop to a radiator [Buford 2011], but this involves additional components, weight and energy consumption for the coolant pump. Another option is utilizing the vehicle's aerodynamics to direct airflow to a heat sink thermally coupled to the battery cells, or to utilize other vehicle components as heat sources or sinks. Little information can be found exploring these topics or their effectiveness in the literature, most likely due to the complexity of the functional integration required. The primary disadvantage of passive thermal management is conditioning the battery when the ambient is far from the optimal operating temperature. As a result, preconditioning the battery is not always possible; however, beginning each trip with a conditioned battery, especially in adverse climates (e.g. winter cold-start) significantly increases vehicle range. In general, passive thermal management alone is used when cell heat loss is low, which is more likely to occur in a BEV because the large system size results in a low current per cell. Furthermore, BEV cells are typically designed to have a higher capacity and lower maximum power

[Sterner 2014b]. PHEVs, which also have an all-electric driving mode like a BEV, draw an equivalent power from a smaller system of high-power cells. The resulting increase in heat-loss per cell increases the risk of premature cell aging. As a result, active thermal management is the state-of-the-art for PHEV applications.

State of the Art of Active Thermal Management

Wiebelt et. al. (2009) describe four basic categorizations for the active thermal management of battery cells: convection (forced air over the cell), conduction to a BTMS on the edge of cells (outside the module), conduction to a BTMS from between cells (spacers), or conduction over the electrical contacts (pole cooling) [Wiebelt 2009]. Pesaran et. al. (2006) also distinguish between direct and indirect cooling at the cell surface by gas (air), liquid (dielectric fluids) and solid (phase change material) [Pesaran 2006].

Patents involving forced air cooling are most prevalent for HEV battery systems, which have the lowest thermal management demands due to the focus on fuel consumption reduction instead of electric driving. These systems often draw cabin air, which is already filtered and mostly at an acceptable temperature. Oil in direct contact with the cells, analogous to transformer cooling, has been shown to be suboptimal for vehicle application due to the high viscosity and thus pumping power required [Pesaran 2003].

Coolants and refrigerants have the potential to remove more energy, but most common choices (water-glycerol, R134a, R1234yf) cannot be used in *direct* contact with the cell casing [Wiebelt 2009, Pesaran 2003]. Therefore, the additional layers between the heat transfer medium are of critical importance. The actual contact interface between a cell and BTMS must be electrically insulated yet thermally conducting. Because of the manufacturing tolerances inherent to both the cells and the BTMS, as well as real surface roughness, areas of poor contact and air gaps can result.

The cell type and size influence thermal management design from a technical standpoint. For example, large pouch cells have thin outer edges resulting from the manufacturing process, often making the use of spacers necessary to achieve a homogeneous temperature distribution; on the other hand, thermally contacting small cells on one side may be adequate [Pesaran 2003]. Therefore, the breadth of patents for thermal management systems for battery cells is quite large, but some common trends can be seen. Only a few representative patents are cited: many more exist. The following is a summary of the state-of-the-art of active thermal management for prismatic Lithium-Ion cells:

- Forced air over the cells and cell terminals using cabin air and various combinations of additional heat exchangers [Carduso 2013, Inuoue 2014, Okawa 2013, Yu 2013];
- Cooling plates are presented for all cell types. The primary focus of the patents is the internal layout [Boddakayala 2013, Kwak 2013, Toepfer 2013] or the production steps and integration with the module [Daubitzer 2012, Stripf 2012, Weileder 2012];
- Spacers are generally considered for pouch cells because of the large contact surface between cells versus the external "seam." In terms of coolant, forced air through a porous spacer has been investigated in the literature [Giuliano 2012], but conducting fins or spacers using liquid coolant are more common [Isermeyer 2010, Syed 2013, Teng 2011];
- Liquid cooling with a water-based mixture and refrigerant (R1234yf or R134a) are the most common coolants in current production PHEVs;
- Alternative techniques using for example heat pipes [Huehner 2012, Walser 2014] or thermo-electric elements [Gawthrop 2013, Song 2013] are less common;
- The state-of-the-art for the thermal management concept for prismatic PHEV cells is the cooling plate.

While the state of the art for prismatic PHEV-format cell thermal management is a cooling plate, variations in the layout, materials, production techniques used, etc. exist, leaving the need to identify a system that is optimized not only for thermal performance, but also for energy consumption, vehicle suitability, production complexity and economic viability.

1.3. Thermal Management Concepts Considered in this Research

Active thermal management concepts are the focus of the feasibility study because of the high thermal loads generated by a high-performance PHEV. Refrigerant, though considered one of the most compact ways to remove energy from the battery pack [Neumeister 2010], is not considered in this research due to existing project boundary conditions and the demand to restrict the use of refrigerant in the vehicle. Based on the state of the art and trans-industry technology scouting, three main concepts and derivations thereof are were selected for the experimental feasibility study:

- Cooling plates with water-glycerol coolant (Section 5);
- Ceramic spacers with water-glycerol coolant (Section 6);
- Heat pipe spacers joined at a collector plate then to a remote heat exchanger via long heat pipes (Section 7).

In order to characterize the thermal management systems analyzed, a classification is established. The classification is based in the thermal contact surface between the BTMS and module as well as the characteristic flow channel dimension. The concepts analyzed in this research are classified in Figure 1.3.

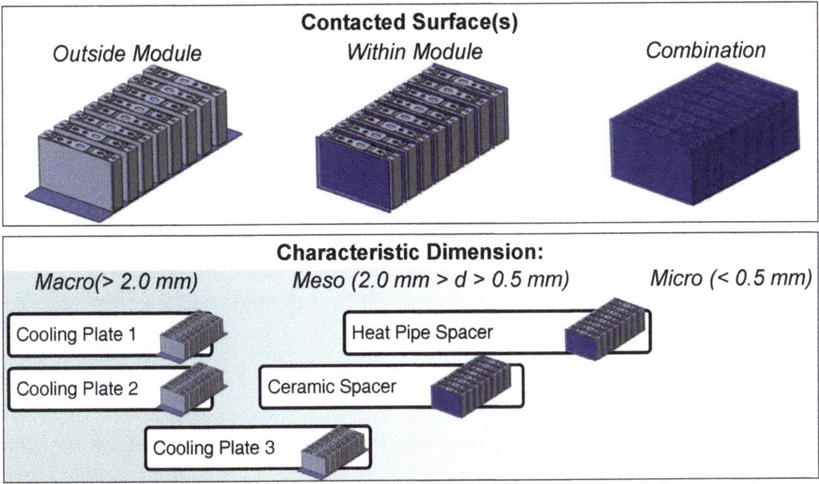

Figure 1.3: Classification of the thermal management concepts analyzed in this research based on the contacted surface of the module (top) and the characteristic flow dimension (bottom).

The motivation behind the selection of each concept, as well as potential advantages and disadvantages, are described in the corresponding sections. The most promising variations of each concept are then compared amongst each other at various use cases as well (Section 8). The reference case for the comparison is a module without thermal management (Section 4), as this is the ideal form in terms of added production complexity and thus cost.

1.4. Goals and Novelty of this Research

This work seeks to enhance the field of thermal management for automotive large-format prismatic Lithium-Ion cells through the following contributions:

- **An experimental method to rapidly and reproducibly analyze new thermal management concepts in early product development states**

  The use of a so-called "Smart Battery Cell" (SBC) is proposed to replace a real Lithium-Ion cell in experimental trials. The SBC contains 16 thermocouples integrated below the cell surface to provide a spatial measurement resolution not currently found in the literature. Because the data acquisition and control systems are integrated within the cell casing, the thermal contact surface is left undisturbed, facilitating reproducible measurements independent of thermal management system layout.

  The heat loss of the cell is reproduced via eight heating zones per cell, distributed over a geometry mimicking the internal layout of a real cell. Thus, the temporal thermal behavior of the real cell can be modeled. The heating elements allow for the analysis of thermal management concepts under the most extreme ambient conditions and cell loads, where the use of a real Lithium-Ion cell would require exorbitant safety measures. Defect cells can be simulated through applied inhomogeneous heat loss profiles to analyze the response of the thermal management system and the effect on other cells

  An internal cooling circuit serves to drastically reduce the reconditioning time, allowing for the completion of 10 to 20 times the experimental trials in a given time period. The rapid reconditioning circuit allows the SBCs to be used as Hardware in the Loop (HiL). The use of the SBC versus a real cell eliminates the cycling effects of real Lithium-Ion cells, which allows for a reduction in the total number of trials, as the variances caused by the active cell chemistry can be accounted for via the cell heat loss model.

  The SBC was developed iteratively by analyzing both the thermal behavior and basic geometry of a prismatic Lithium-Ion cell.

- **A technical feasibility study and multifactorial analysis of diverse thermal management concepts**

  The thermal management concepts analyzed in this research were selected based on the utilized materials and an anticipated ease of assembly under the hypothesis that a production-friendly system will result in lower costs. The thermal management concepts and variations thereof are analyzed at relevant operating conditions and use cases, such as cold weather operation, cold start performance, rapid charging, and high temperature operation. The concepts are analyzed at a module-level, which provides information regarding inhomogeneous cell aging not possible at the cell level. Detailed experimental analyses in the literature do not often consider more than three to four cells, where as in this work, eight cells are combined.

Based on the experimental data and in collaboration with component manufacturers, the thermal performance, energy consumption, vehicle suitability, production, and cost of the thermal management systems are analyzed using unique and existing criteria. From the presented multifactorial analysis, a system and/or characteristics thereof can be selected based on the requirements of a specific vehicle project.

Furthermore, mathematical models of the average module temperature based on the first law of thermodynamics are generated for each concept.

- **Recommendation for the ideal system for a high performance PHEV**

  Novel thermal management concepts, the state of the art thermal management system for prismatic large-format cells (a cooling plate), and a system without thermal management are analyzed. From the multifactorial analyses, recommendations for the ideal thermal management system for a high performance PHEV using large-format prismatic cells are presented.

1.5. Structure of this Research

This research contains two main parts: the presentation of the methodology and the analysis of thermal management concepts.

Methodology

Section 2 presents the various criteria derived and adapted for the analysis of thermal management concepts.

Section 3 presents the SBC as a novel method to experimentally analyze battery thermal management systems. The SBC models the thermal behavior in hardware form, providing an alternative to simulation. The analysis of the actual cell as well as the design and validation of the final SBC design are presented.

Analysis

Section 4 presents an eight-cell module with no thermal management, which serves as the reference system for the multifactorial analysis in Sections 5 through 8.

In Section 5, three different cooling plates are analyzed in order to draw conclusions regarding the ideal cooling plate design based on the evaluation criteria presented in Section 2.

In Section 6, active spacers (between the cells) are analyzed in various configurations. The optimization of spacer geometries and materials is analyzed numerically. The results are evaluated based on the criteria presented in Section 2.

In Section 7, a two-part demonstrator consisting of both heat pipe spacers and long heat pipes is analyzed. The spacer position is varied as in Section 6. The results are analyzed based on the evaluation criteria presented in Section 2.

The most promising thermal management concepts analyzed in Sections 5 through 7 are compared in Section 8, resulting in a recommendation for the thermal management system of a high performance PHEV.

Conclusion

Section 9 provides a summary of the work, including suggestions for improvement and further applications of this research.

Methodology

2. Evaluation Criteria

In order to evaluate thermal management systems for a luxury, high performance PHEV at various use cases, evaluation criteria are determined. For an efficient multi-factorial analysis of thermal management systems, only the most relevant criteria must be considered. Therefore, five criteria, categorized as quantitative or qualitative, have been selected for this research:

- Quantitative:
  - Thermal performance
  - Hydraulic work
  - Vehicle suitability
- Qualitative:
  - Producibility
  - Economic viability

The quantitative criteria are experimentally measured, while the qualitative criteria are derived from expert opinions and evaluated relative to one another, as they are more abstract and cannot be directly measured. Access to specialists in the analyzed fields of technology makes the conclusions drawn from the qualitative criteria valuable as well. Because of the interdependence of production complexity and economic viability, these two criteria are combined. The derivation and explanation of the general evaluation criteria utilized for the analysis of all thermal management systems are presented in this section. Classifications or analyses unique to a certain concept are presented in the corresponding section (Sections 5 through 7). The evaluation criteria are all applied at a module-level (Figure 1.2) so that they can be universally applied to various system layouts.

2.1. Thermal Performance

This research is based on a production-oriented analysis with the primary goal of reducing system costs by minimizing production complexity. Additional components not critical to cell function and safety should be minimized. For this reason, the reference module for this analysis of thermal management systems is a module with no thermal management (presented in detail in Section 4). The purpose of thermal management, as shown in Section 1, is to minimize temperature-induced aging.

In this research, the evaluation criterion used to determine temperature-induced aging is "thermal performance."

The criterion thermal performance consists of three components:

- the average module temperature $\bar{T}_{module}$
- the differences in the average temperatures of the individual cells $\Delta\bar{T}_{cells}$
- the differences between the temperature gradients over individual cells $\Delta(\Delta T_{cells})$.

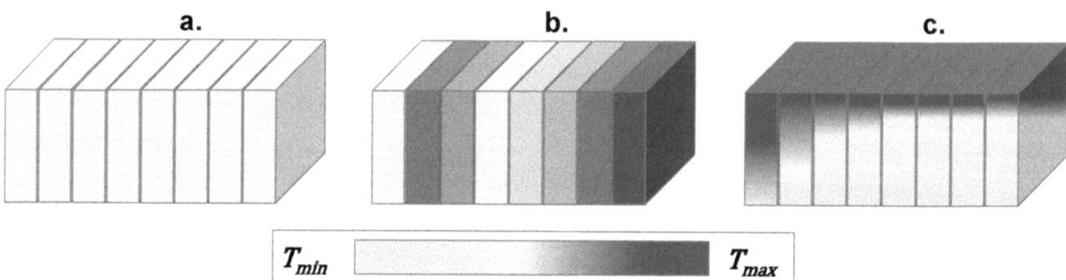

Figure 2.1: The eight-cell reference module schematically showing the three components of the criteria for temperature-induced aging "thermal performance": the average module temperature $\bar{T}_{module}$ (a), the differences in the average temperatures of the individual cells $\Delta\bar{T}_{cells}$ (b), and the differences in the magnitude of the temperature gradients over individual cells $\Delta(\Delta T_{cells})$ (c).

The differences in the three aforementioned quantities are shown schematically in Figure 2.1 and described in the following passages.

Average Module Temperature $\bar{T}_{module}$

The average temperature of the module $\bar{T}_{module}$ provides an indication of the general temperature level of the system. The instantaneous average temperature of the 128 thermocouples is plotted over time, and is used to approximate a solution to the first law of thermodynamics assuming the module as a lumped mass:

$$C\frac{dT}{dt} = \dot{Q}_{loss} - hA_\infty(\bar{T}_{module} - T_\infty) - \dot{Q}_{BMTS} \tag{2.1}$$

Where C is the thermal capacity of the module, $\frac{dT}{dt}$ the temperature change as a function of time, and $\dot{Q}_{loss}$ the irreversible heat loss due to the cell internal resistance. The term $hA_\infty(\bar{T}_{module} - T_\infty)$ is the heat transfer to/from ambient and $\dot{Q}_{BMTS}$ the heat transfer to/from the BTMS. For the case of a liquid single phase coolant, the heat transfer between the coolant and the BTMS is:

$$\dot{Q}_{BMTS} = hA_{BTMS}(\bar{T}_{module} - T_W) \tag{2.2}$$

In this way, the fitted values of thermal conductance hA_{BTMS} (W^1K^{-1}) of various BTMS can be compared to one another and to the reference module. The generated models can then be used to predict the behavior of the thermal management systems under other use conditions.

Differences in the Average Temperatures of the Individual Cells $\Delta \bar{T}_{cells}$

The differences in the average temperatures of the individual cells within the module $\Delta \bar{T}_{cells}$ indicate if the cells are exposed to a homogeneous thermal environment and thus if they will age at similar rates. When a single cell prematurely ages, the complete battery system is affected. Cells are electrically contacted in series to raise the system voltage, and in parallel to raise the system capacity. When connected in series, the "chain" formed is susceptible to the failure of a single cell. Aging of a single cell additionally lowers the total system voltage, and thereby the vehicle range [Koehler 2013, Kampker 2014]. Additionally, when chains of cells connected in series are connected together in parallel, each cell voltage must be monitored to ensure an equivalent voltage [Koehler 2013]. As a result, the voltage of each chain of cells connected in series must be balanced to an equivalent value. When a single cell ages prematurely, capacity losses for the whole system occur, resulting in reduced vehicle range.

In this research, $\Delta \bar{T}_{cells}$ is defined as the sample standard deviation between the highest average temperature of an individual cell ($T_{n,max}$) and the lowest average temperature of an individual cell ($T_{n,min}$) over the eight cells in the module:

$$\Delta \bar{T}_{cells}(t) = \sqrt{\frac{1}{7}(\textstyle\sum_{n=1}^{8}(T_n - \bar{T})^2)} \tag{2.3}$$

The temperature difference over the module is plotted as a function of time in the analysis, and the time averaged value is utilized in tables.

Differences between the Temperature Gradients over Individual Cells $\Delta(\Delta T_{cells})$

The difference in the magnitude of the temperature gradients over each individual cell $\Delta(\Delta T_{cells})$ is another indication of inhomogeneous aging used in this research. This term compares the temperature gradients over each individual cell ΔT_{cell} to one another, thereby expressing the susceptibility of the cells in the module to age at different rates. The standard deviation of the temperature is again used instead of purely the difference between the maximum and minimum temperature in order to mitigate the influence of data outliers from the 128 measurement points throughout the module. Furthermore, the SBC contains 16 measurement points on the cell jelly-roll: because this is a small percentage of the total volume, the sample standard deviation (Equation 2.4) at each time (t) is used, which corrects for the fact that the true jelly roll mean temperature is not known [Navidi 2006].

$$\Delta T_{cell} = \sqrt{\frac{1}{15}(\textstyle\sum_{n=1}^{16}(T_n - \bar{T})^2)} \tag{2.4}$$

To compute $\Delta(\Delta T_{cells})$, the difference between the maximum and minimum ΔT_{cell} of an individual cell in the module is taken:

$$\Delta(\Delta T_{cells})(t) = \Delta T_{cell,max}(t) - \Delta T_{cell,min}(t) \tag{2.5}$$

Plots of $\Delta(\Delta T_{cells})$ as a function of time are shown in the analysis, and the time averaged value is utilized in tables.

2.2. Hydraulic Work

The theoretical hydraulic work required for the flow of a single-phase coolant through a thermal management system is used as an indication of energy consumption. The energy required for a thermal management concept is important because of the effect on vehicle range: the more energy consumed by the BTMS and more generally the vehicle thermal management system, the lower the vehicle range.

In this research the theoretical pumping power between use cases and concepts is compared to identify energy efficient thermal management concepts. The theoretical pumping power (P), defined by:

$$P = \dot{V}\,\Delta p \tag{2.6}$$

is a function of the volumetric flow rate of the coolant ($\dot{V}$) and the pressure drop over the circuit (Δp). The pump efficiency is omitted since only the comparison is of importance. As such, a lower calculated pump power indicates an energy savings. Equation 2.5 thereby considers both the contribution of the BTMS flow channel layout/geometry (Δp) and the required operating conditions ($\dot{V}$) to the theoretical pump work.

As technology demonstrators, the experimental setups are not optimized to minimize hydraulic work. Therefore, investigations into the optimization of certain design aspects were performed using computational fluid dynamics. The lessons learned from these analyses are presented and discussed in the corresponding sections.

Analog to the analysis of thermal performance, the reference for the comparison of required hydraulic work is a battery module with no thermal management, which consumes no additional energy. Parameters influenced by the total vehicle thermal management system are not considered: this research is intended as an input for such modeling.

2.3. Vehicle Suitability

Vehicle suitability is evaluated in terms of BTMS weight and size. BTMS leak-tightness is also important if a coolant is used. This characteristic was ensured prior to the experiments: a leak-prone system was not analyzed further. Additional characteristics such as mechanical stability (vibrations, crash, etc.) and chemical resistance (corrosion, system lifetime) were not considered in this research, but these criteria must be considered under the boundary conditions of the final battery system and vehicle application.

The available space in the vehicle is a critical factor to BTMS selection, especially in a PHEV, where both a conventional- and electric drive train are integrated within the same vehicle. Depending on the vehicle layout, a specific dimension may be more critical than another: for example, in an under-floor battery, height (z-direction) may be more important than the system length or width. Additional vehicle weight detracts from vehicle range. Considering the current high cost of battery systems, reducing vehicle weight instead of adding cells may in some cases be a more economical method of increasing vehicle range. Weight, like size, is evaluated in comparison to the reference case (a module without a BTMS). Therefore, the dimensions of a module (L_i) with a BTMS is considered in all three Cartesian directions ($i = x, y, z$), and in terms of total volume (V) and weight (m). These values are compared to the size and weight of the reference battery module without a BTMS by the following formulae:

$$\Delta L_i = \left(\frac{L_{BTMS} - L_{ref}}{L_{ref}}\right)_i \tag{2.7}$$

$$\Delta V = \left(\frac{V_{BTMS} - V_{ref}}{V_{ref}}\right) \tag{2.8}$$

$$\Delta m = \left(\frac{m_{BTMS} - m_{ref}}{m_{ref}}\right) \tag{2.9}$$

where ΔL_i, ΔV, and Δm are the difference in length, volume, and mass related to the reference case.

The technology demonstrators used in the experiments served as a feasibility study, and as such are not optimized to the minimal size and weight possible with the analyzed production technology. Therefore, in conjunction with component experts, projections for future component sizes and weights were analyzed and are presented in the corresponding analyses in Sections 5 through 7.

2.4. Producibility and Economic Viability

Traditionally, product design is followed by production planning based on, for example, cost point, output targets, company goals, or market strategy [Guenther 2012]. This research proposes considering economical or efficient production techniques in *advance* of product design in order to speed development and lower production costs. Additionally, under the assumption that vehicle electrification should reduce not only tail pipe emissions but also global emissions, the total product lifecycle is considered conceptually.

The presented analyses of the producibilty is performed qualitatively, because the detailed planning of the complete production process for a single product is outside the scope of this research. Therefore, industry experts and the literature are utilized to compare the producibility of the considered production techniques.

The importance of a well-conceived production process is shown through the calculation of the production yield of a process consisting of multiple steps: based on a 15-step production process, with each step having a net-yield of 95 to 99% (acceptable values), the total yield of on-spec parts for the complete process varies from 46 to 86% [Pettinger 2013]. The production process must also be optimized for the specific product, as many processes are only economical at a certain volume [Grote 2009]. One method of checking the harmony between a product and the production process is shown in Figure 2.2. This methodology was implemented in the analysis of the technology demonstrators: in the prototype phase, all production steps were chosen accordingly as to realize a balanced production. In the analysis of future mass production concepts, all steps were then scaled together, as shown in Figure 2.2.

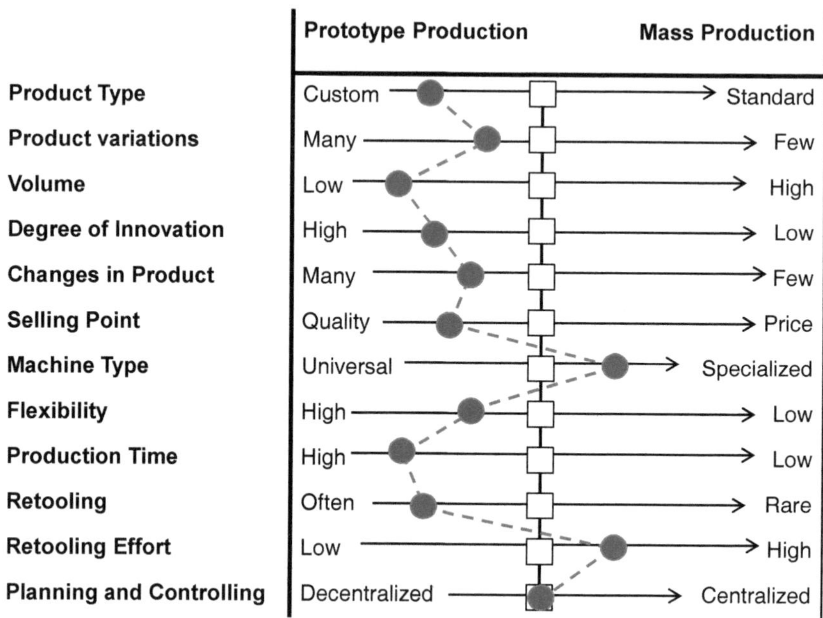

Figure 2.2: The general process for determining a balanced production process. Shown are a well-balanced (square) and a poorly balanced (circle) production process for a product. Adapted and translated from [Guenther 2012].

If electrification of vehicles is the future method of transportation, the battery system must certainly be laid out for mass production. Using the module-based approach of construction a battery system (Section 1), the volume of like-parts can be raised further. With respect to Figure 2.2, a balanced production process should lean toward the characteristics at the right of the diagram.

In the current research, however, the technology demonstrators for the feasibility studies of thermal management concepts are prototypes. Small-scale processes are used to check the technical feasibility of the designs: the design and assembly of these prototypes provided invaluable insight into potential challenges for future designs. Therefore, the lessons learned from this phase of the research are discussed and utilized in order to recommend an improved product and production process for an eventual mass production. In order to provide a technical background for the evaluation of production balance, the categories of manufacturing as classified by the standard DIN 8580 and shown in Figure 2.3.

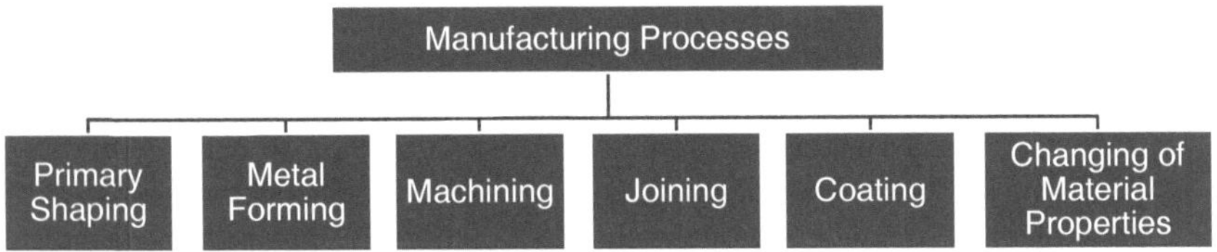

Figure 2.3: The categories of fabrication according to DIN 8580 [DIN 8580].

The classifications are discussed briefly in the following as an overview, but topics specific to certain technology demonstrators are covered in depth in Sections 5 to 7.

- **Primary Shaping**

  Primary shaping covers broadly the formation objects from a material in bulk or stock form, rather by casting, injection molding, or sintering [DIN 8580]. A broad variety of materials may be used, and material properties can be widely influence by the process itself, for example by the mold or the means of material distribution [Grote 2009]. The type of primary shaping method used influences final part quality and can even eliminate additional production steps such as machining [Grote 2009]. One of the promising manufacturing technologies of the future, generally referred to as 3D-Printing, belongs in this classification. Casting, especially near-net shape casting, is superior to machining in respect to material waste and environmental efficiency.

- **Metal Forming**

  Metal forming is defined as controlled plastic straining to a desired shape through bending, shearing, tension, compression or tension and compression [DIN 8580]. The actual forming techniques can be grouped into bulk forming (forging, drawing, rolling, upsetting, etc.) and sheet forming (deep drawing, hydroforming, etc.) [Grote 2009]. Metal forming processes also have a high material utilization and energy conservation, as well as short production times as well as material properties ideal for dynamic loading; on the other hand, higher production output is required to counter the high investment costs and geometry can be limited in some cases [Grote 2009].

- **Machining**

  Machining is the removal of material by mechanical or non-mechanical means, which includes common processes such as cutting, milling, sanding as well as surface preparation/cleaning [Grote 2009, DIN 8580]. Machining is often performed in combination with other production techniques, but as an additional production step the total production time and complexity are increased. Machining techniques have also been used to create many of the experimental technology demonstrators because of the flexibility provided for prototype manufacturing.

- **Joining**

  When a component, such as a thermal management system, cannot be made from one part, a joining process is required. The scope of joining technologies is tremendous: over eighty different techniques are used to varying degrees worldwide [Grote 2009]. The technology demonstrators presented in this research utilize a wide variety of joining techniques, which are discussed in depth in Sections 5 through 7.

- **Coating**

  A coating can be applied to protect or alter the surface of a part, or prepare it for another process such as adhesion [Ebnesajjad 2011]. Common applications include anti-corrosion coating, painting or chroming. In relation to BTMS, a coating can also provide electrical insulation or enhance the thermal contact between a cell and BTMS.

- **Alteration of Material Properties**

  Material properties are commonly altered by mechanical forming (rolling, etc.), heat treating, or mechanical-thermal processes [Grote 2009]. Using such techniques, the fundamental material structure ins altered, and mechanical behavior can be influenced. The effect of cold-rolling, which alters material properties, is investigated in conjunction with two joining techniques in Section 5.

Various manufacturing techniques and combinations thereof were considered in this research. With this overview of production techniques, the basics behind the qualitative analysis of production complexity can be understood. Certainly more combinations and applications can be generated by experts. The discussions of the technology demonstrator production are intended to give a high-level recommendation of the direction to be followed, and not a detailed manufacturing analysis. Utilizing this technical basis and the concept of a balanced production process, the technology demonstrators are evaluated for their current state and for their potential in a mass production process.

Additionally, the total life cycle emissions of the product must be lowered in order to maximize the environmental benefit of electrified vehicles. Emissions during the useful lifetime of an electric vehicle are effected by the energy efficiency of the vehicle and the composition of the electricity used for charging. During this phase, the thermal management must guarantee the thermal state of the cells, use little energy, and weigh as little as possible, while the vehicle operator must charge from renewable energy sources. However, a complete life cycle analysis must also consider at least the raw material production, manufacturing, and the end of life (i.e. disposal or reuse) [Marretta 2012]. Depending on the raw materials, the energy required during this stage can greatly outweigh the actual vehicle use. One method of reducing the impact of the material production process is through the use of recycle materials: using steel scrap versus primary steel requires approximately 1/3 the energy input, while the use of scrap aluminum reduces energy consumption over 20 times with respect to primary aluminum; however, the mechanical properties are influenced due to the accumulation of nickel and copper (steel) or silicon (aluminum) [El-Mahallawi 2013, Yellishettya 2011]. As such, to maximize the potential of aluminum in terms of global emissions, the manufacturing process as well as the component itself should be designed to use recycled materials [Marretta 2012]. The interplay between the energy consumption of the various life cycle stages is shown schematically in Figure 4.

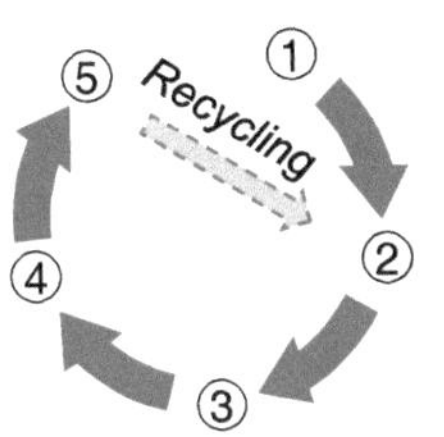

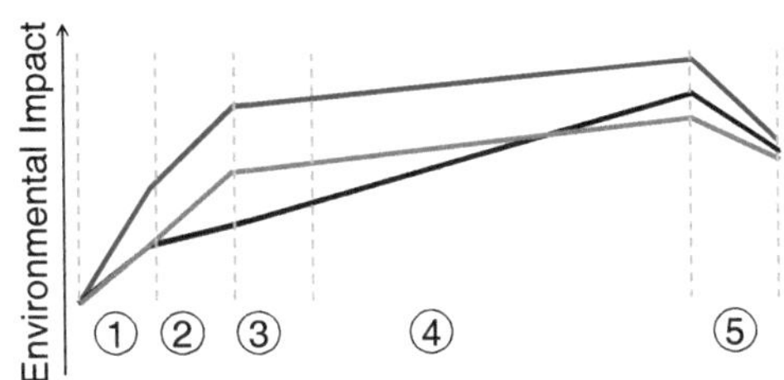

Figure 2.4: Left, a schematic of the life cycle of a product beginning with the material extraction (1), the material processing (2), the manufacturing process (3), the useful lifetime of the product (4) and the end of life of the product (5). Right, the same life cycle phases shown over an arbitrary time scale versus the environmental impact. Adapted from [Remmen 2007] and [Krinke 2011].

The exact emissions due to material processing and component manufacturing are dependent on the energy source. In order to compare the materials equally, the estimated energy consumption per ton of material produced is shown in Table 2.1.

Table 2.1: Comparison of the energy required for the production of various materials used in the technology demonstrators in this research [El-Mahallawi 2013, IEA 2007, Yellishettya 2011].

Material	Material Production without Recycling (MJ per ton)
Steel	20 – 31
Aluminum	50 – 150
Copper	30 – 40
Ceramic	9 – 30

As shown in Table 2.1, the energy consumption of the manufacturing process can be significant. This is especially true in the case of electric vehicles using Lithium-Ion cells, where the extraction and processing of rare-earth minerals is required. The relationship between energy usage over various life cycle stages is shown in Figure 2.5.

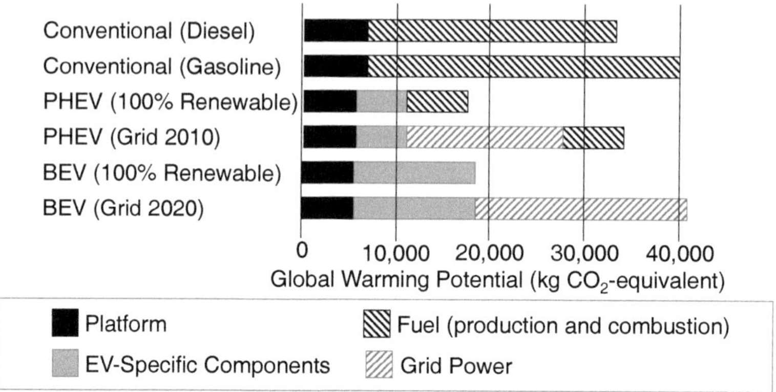

Figure 2.5: Comparison of the contributions to life cycle emissions of the manufacturing stage ("Platform" and "EV-Specific Components") to the emissions during use for conventional vehicles as well as electrified vehicles powered with and without renewable energy sources. Adapted from [Held 2011].

Figure 2.5 shows the relative magnitude of the energy required during the material extraction and production process in relationship to the actual energy consumption during vehicle use. Clearly, all life cycle stages must be considered to maximize the environmental benefit of electrified vehicles.

The economic viability of the technology demonstrators is not reflective of potential production costs in the future. Therefore, in conjunction with component suppliers, manufacturing and part costs are estimated and presented in relation to other concepts. The primary goal of this evaluation criterion is to eliminate technologies that are inherently expensive, rather it be the material used or the production process. Both the investment and the component costs are considered, and the analysis is performed from the viewpoint of an automotive OEM. Therefore, a specialized part (e.g. extruded ceramic) is assumed to be sourced, with results in lower investment costs but higher part costs.

2.5. Summary

The presented analysis techniques are utilized throughout this research to evaluate the concepts and facilitate in the design of future BTMS. The individual analyses comparing variations of a specific thermal management concept are shown in Sections 5 through 7, while a global comparison amongst concepts is presented in Section 8.

3.   Experimental Method

This hardware-based thermal battery cell model (SBC) of a prismatic PHEV-2 cell has been developed to enhance the analysis of battery thermal management systems (BTMS) beyond traditional experimental methods or simulative approaches. The general goals of the presented method are to:

- Temperature measurements at a spatial resolution not currently found in the literature, without disturbing the thermal contact surface;
- Minimize the amount of experimental trials and time;
- Guarantee reproducible conditions and a direct comparison of systems;
- Facilitate scalability and adaptability (accommodate changing cell behavior during the development process);
- Test thermal management systems in early stages of development safely.

These goals are achieved through the development of a "Smart Battery Cell" (SBC). The steps in the development of the SBC are presented on the basis of a DIN Standard 91252 prismatic PHEV-2 cell, but the method can be implemented for other cell types. Existing methods of analyzing thermal management concepts and simultaneously the resulting effect at the cell do not facilitate and efficient experimental comparison. These shortcomings are discussed in further detail in Section 3.1. Thereafter, the creation and validation of the SBC are presented.

3.1.   State of the Art

A complete experimental analysis of battery thermal management systems requires both a detailed analysis of the *effect on the battery* and of the *thermal management concept performance*.

In order to measure the effect on the battery, the temperature resulting at the cell as a result of the thermal management system must be determined. In order to compare various systems, it is desirable to repeatedly measure the temperature at the same location and under the same boundary conditions. Comparing thermal management systems via simulation models eliminates the cost and infrastructure of an experimental setup; however, the accuracy of a simulation depends on the accuracy of the inputs. For example, a large source of error is the thermal contact interface, which is difficult to calculate. The effect of the thermal contact interface can however be isolated and observed in experimental trials. For this reason, a hardware-based thermal model of the cell is chosen.

Most analyses of cell characteristics or behavior found in the literature use actual cells. Arguably the simplest and most prevalent method for measuring the effect of a thermal management system on battery cells is by attaching thermocouples at various positions over the cell casing. Countless experiments are performed in this way, as thermocouples of varying size and accuracy are commercially available. However, some thermal management concepts include small gaps between cells and various components, making the integration of thermocouples at all desired locations difficult. The required electrical insulation between the thermocouple and the cell can affect thermal contact surfaces, resulting in altered behavior. The placement of a thermocouple in the exact same position on a cell in diverse thermal management concepts may not be possible due to interference with other components. Essentially, thermocouples themselves are not the problem, but rather their use on the exterior of the cell. Additional experimental methods considered for the comparison of the effect of thermal management systems on battery cells are briefly discussed below.

- Thermal imaging is a useful non-invasive technique that has been effectively implemented for single cells and groups of cells [Pesaran 2013, Giuliano 2012, Robinson 2013]; however, the surface analyzed must be visible to the camera, meaning that temperatures and gradients for example directly between two cells or between cells and BTMS cannot be measured.

- In-situ temperature measurements using thermocouples or temperature sensors within the cell provide data at the critical chemically active layers, allowing local temperatures to be measured. This technique has been applied on numerous cell types [Heubner 2013, Li 2013, Mutyala 2014, Yang 2013, Zhao 2012]; however, using actual cells carries the increased cost of additional experimental trials due to the variance in cell behavior that occurs during cycling. Additionally, the presence of thermocouples or similar within the active chemical layers can influence cell behavior. Actual cells also present safety hazards (e.g. because of coolant leakage) with new, unrefined thermal management concepts.

- Calorimetry has been used to determine the heat loss of a cell (or cells) very precisely under defined loads, and to then approximate the heat transfer coefficient and cell temperature using a lumped capacitance method [Pesaran 2013, Hallaj 2000]. This method is useful for determining for example total energy removal, but cannot isolate local gradients and discrepancies between- and across individual cells that lead to premature aging.

- Electrochemical impedance spectroscopy measurements have been used to determine the internal temperature of actual cells [Robinson 2013, Raijmakers 2013, Schmidt 2013]. While temperature differences between cells can be measured, gradients over the individual cells and local hotspots induced by a thermal management system cannot be localized. This technique is however very promising for battery system monitoring.

- Wavelength based methods involving emission, absorption and scattering [Childs 2003] to determine thermal behavior are not considered, as they are not available to the authors.

- Various combinations of any of the aforementioned methods, though not discussed, are present in the literature.

Regardless of the measurement technique, a typical experiment using actual battery cells is time-consuming: the high thermal mass of a battery module or system requires a large reconditioning time to ensure constant initial conditions. As a result, often only one trial per day can be performed. Additionally, the temperature sensitivity of the

Lithium-Ion cells results in varying thermal behavior depending on the travel, use, and storage history of the cell. To compare *only* the effect of thermal management concepts, many additional trials are required to eliminate the statistical variance caused by each cell's individual history. By reducing experimental time, more trials can be performed more quickly so that product development can be influenced at an early stage, thereby reducing time to market and development costs. Experimental trials with large-format Lithium-Ion cells in early stages of BTMS development are dangerous. For example, compromising the electrical insulation or inducing high temperatures in any way can result in severe safety risks: not only is the electrical energy in large format cells high, but far more the chemical energy released during failure proposes a significant safety hazard. Coolant leakage or BTMS failure are realistic risks during prototype testing, and as a result specially qualified technicians and safety equipment are required. Furthermore, testing the robustness of a BTMS by simulating a damaged or prematurely aged cell is also difficult to perform reproducibly, as bringing a cell to a pre-determined aging state is difficult. By avoiding the use of actual cells, experimental safety and reproducibility can be realized far more quickly and efficiently.

Alternatives to actual cells are presented in the literature primarily for the analysis of the performance a specific thermal management concept [Tran 2013, Wang 2014, Belyaev 2014, Lukhanin 2012, Lukhanin 2013]. In these cases, only two to three cells are considered. This small-scale does not accurately capture the effect of a thermal management concept on a module, or more precisely the susceptibility to inhomogeneous aging.

A simple heat source, such as an electric heater, is commonly used. More accurate and validated hardware-based cell models are rare in comparison. Wang et. al. [2014] use a casing filled with Atonal 324 to simulate the thermal capacity of the actual cell analyzed. Belyaev et. al. [2014] and Lukhanin et. al. [2012, 2013] utilize a heating plate and various layers to simulate the thermal properties of a pouch cell; however, the temperature sensors are included on the surface, which disturbs the thermal contact interface, and rapid reconditioning is not possible between trials.

No models having a high spatial density of temperature sensors integrated within the casing or using an active reconditioning system are known to the author. The use of integrated temperature measurement (not on the cell surface) as well as rapid reconditioning are significant arguments for a hardware-based model versus an actual cell. Furthermore, combining multiple such hardware-models allows for a level of spatial resolution not currently available in the literature.

In the following Section, the analysis of a real Lithium-Ion cell is presented as the basis for the methodology developed within the scope of this research.

3.2. Analysis of a Prismatic Cell

Competency and know-how for the production of high-capacity Lithium-Ion cells is currently not a focus of automotive original equipment manufacturers (OEM). Therefore, from the perspective of an automotive OEM, sourcing a battery cell is currently the preferred strategy. To maintain warranty coverage, the specifications given by the cell manufacturer must be upheld: it must therefore be guaranteed that new thermal management concepts meet these requirements. As stated in Section 3.1, the use of an actual cell is not always advantageous in determining the effect of thermal management concepts on the cell. Therefore, it is the premise of this research that the hardware-based cell model, or SBC, must dynamically replicate both the average temperature and the temperature gradients at the casing in order to serve as a thermal model of an actual cell.

In order to accurately model actual cell behavior, a series of experiments are performed in the following Section. To provide a novel hardware-based model beyond the state of the art, additional features are integrated into the SBC as described in Section 3.3 and the thermal behavior of the SBC is validated.

Experimental Procedure

The goal of the experimental analysis is to investigate cell thermal behavior at the casing, under the premise that an automotive OEM will source and not produce battery cells. Because the size of the PHEV-2 cell considered, both the average temperature and the temperature distribution are considered. Additionally, because of the dynamic nature of a battery system in operation, a stationary analysis is insufficient. The underlying assumption of the analysis is that if the SBC can dynamically match the thermal behavior of the actual cell in various test conditions, it will also exhibit sufficiently identical behavior to the actual cell when combined to a module or when integrated with a BTMS.

The experimental setup is focused on the exterior of the cell: 40 Type-K NiCr-Ni thermocouples are utilized to measure the temperature (and distribution thereof) over the cell casing: 24 thermocouples are fitted onto the large "front" face of the cell, four on both "sides", five on the "bottom", one near each terminal, as shown in Figure 3.1.

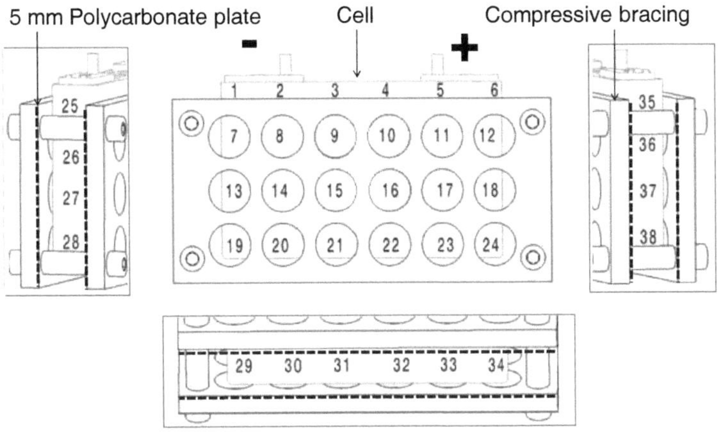

Figure 3.1: Position of the thermocouples over the surface of the cell (numbered one through 38, plus one near each terminal). The modified compressive bracing required for safety is also shown [Schneider 2013].

The other large "back" face of the cell is coated matte black, and has no thermocouples so that thermal imaging can be used to investigate the temperature distribution as well. Unfortunately, the cell requires a bracing for safety during operation and storage because of the generated forces and resulting damage to the cell's active layers. The required safety bracing has however been redesigned to accommodate as many thermocouples as possible and facilitate the use of thermal imaging. Between the bracing and the cell is a 5mm polycarbonate plate to thermally and electrically insulate the bracing from the cell. This insulation serves to mitigate the influence of the bracing on the cell; this is verified via thermal imaging.

The temperature is logged over a CAN-Bus to a PC that is time-synchronized with the high-voltage (HV) testing equipment. The HV-power source is connected to the cell terminals via over-dimensioned cables, as a small cable diameter would disturb the temperature profile of the cell due to the joule heating in the cable. The entire setup is placed in the middle of a 200 Liter climate chamber with temperature and humidity control. Because the climate chamber regulates temperature via a fan, an inconsistent airflow and therefore changing coefficient of heat transfer could influence the average temperature and the temperature distribution, and inhibit the comparison between trials. Therefore, the tests are performed in wind-still conditions at constant relative humidity (40%). To facilitate the wind still condition reproducibly, the chamber and experimental setup are climatized over night to ensure that the entire cell, and not just the casing, are at the same temperature. At the start of the test, when the battery cell cycling begins, the active climitazation is turned off and remains so for the duration of the test.

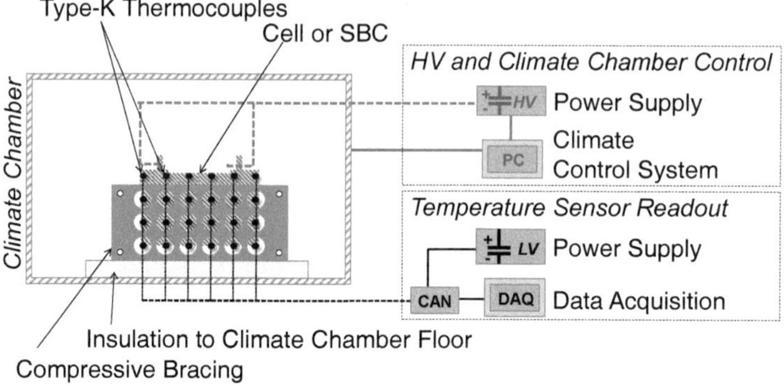

Figure 3.2: Experimental set-up used for the analysis of the behavior of the real cell and the validation of SBC behavior. The thermal imaging camera (not shown) faces the opposing side of the cell. The high-voltage (HV) power supply is only utilized for the real cell.

Additionally, the large volume of the climate chamber in comparison to the cell size means air temperature differences over the test time are minimal. The same natural convection boundary condition is implemented for the module-level comparisons (Section 5 and 6), but due to the size of the experimental setup, a specially designed climate chamber is used (Section 4.2).

Each experiment begins with a fully charged cell (100% SOC, 4.1 V). The cell is the discharged to 20% SOC, and recharged to 100% SOC repeatedly for a total duration of four hours. This process is repeated at various rates C-rates (1, 2, 3, 4, 5 and 6 C), where "C" indicates the capacity of the cell. Thus, for the considered 25 Ah cell, a charging rate of 1 C is equal to 25 A, while a rate of 2 C is equal to 50 A. As a result, more cycles are achieved in the four-hour test time at higher charge/discharge rates. After four hours of cycling, the cells are left in the wind-still climate chamber in order to observe the cool-down behavior.

Five new cells were tested three times at each charge/discharge rate to reduce the effect of variances in cell thermal behavior, which is influenced by the travel, use, and storage history of the cell [Liaw 2003, Rao 2011, Vetter 2005]. The temperature gradients are measured throughout the test via the thermocouples (on five of the sides) and thermal imaging. Contour plots of the measured temperature and the thermal images are used to analyze the temperature distribution, while the measured temperature alone is used to determine the average cell temperature. This analysis was performed within the scope of a closely supervised masters thesis [Schneider 2013]. The same experimental setup, with the absence of the HV connection, is used to validate SBC performance. In this case, the data acquisition, power, and hydraulic connections of the SBC are utilized.

Results and Discussion

From the experimental analysis of the actual cell, a statistically significant temperature distribution across the casing cannot be determined. The measured cell surface temperatures are all within the standard error tolerance of Type-K NiCr-Ni Class A thermocouples [Bernhard 2014, IEC 2007], as shown in Figure 3.3.

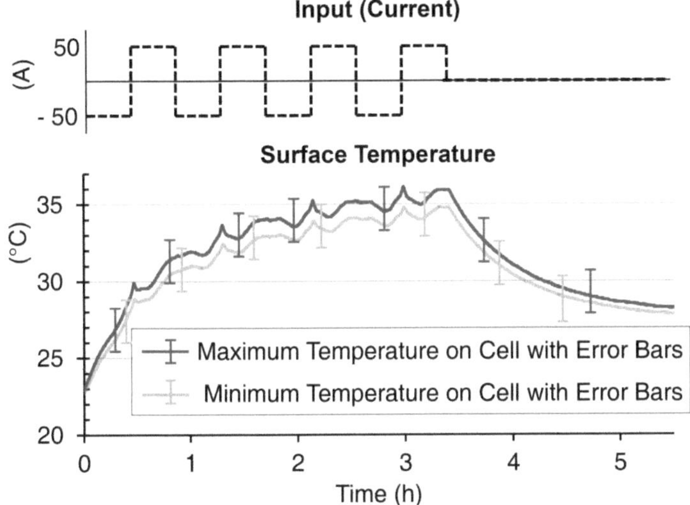

Figure 3.3: The input current (top), followed by the highest and lowest values of measured surface temperature on the real Lithium-Ion cell during 2C charge/discharge cycles [Schneider 2013]. The x-axis is the same for both plots. The maximum temperature observed was generally near the middle of the front surface (Thermocouple 9-10, 15-16, 21-22 in Figure 3.1), while the minimum temperature occurred near the corners. Shown are the error bars for the measurement uncertainty of ± 1.5 K for Type-K NiCr-Ni thermocouples [Bernhard 2014, IEC 2007].

The relatively homogenous surface temperature of the cell can be explained by the internal layout of the cell. The primary component of the cell, the so-called jelly roll, is where the electrochemical reaction occurs. In the PHEV-2 cell analyzed, the jelly roll is created by rolling a large sheet consisting of three primary components: a cathode on an aluminum film, a separator, and an anode on a copper film (i, ii, and iii in Figure 3.4, see Section 1.1. for more details regarding the active chemistry). The aluminum and copper films are drawn together at opposing ends of the jelly roll, and fastened via a current collector tab (1 and 3 in Figure 3.4) that connects to the cell terminals.

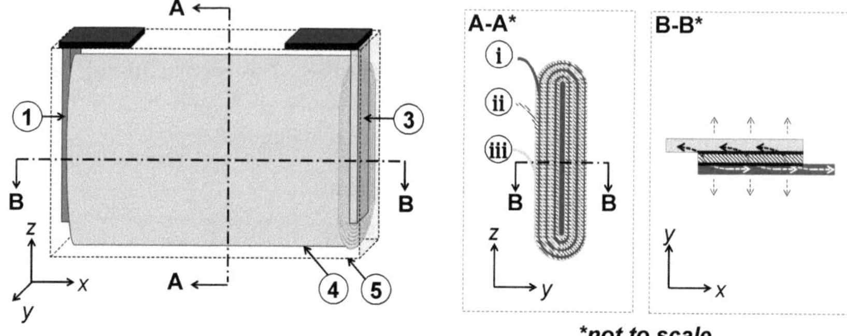

Figure 3.4: Schematic of the internal layout of the cell. Left, the isometric view of the cell showing the copper (1) and aluminum (3) current collector tabs, the jelly roll (4) and the cell casing (5). Section A-A shows the jelly roll shape and the anode with copper film (i), the separator (ii), and the cathode with aluminum film (iii). Section B-B shows a single anode-separator-cathode layer including the paths of conduction within the layers and through the layers. Layer lengths and thickness are not to scale.

The polymer separator has a much smaller specific thermal conductivity than the aluminum and copper films, meaning that the dominant path of thermal conduction is along the copper and aluminum films to the current collector tabs (Figure 3.5). The temperature increase in each the aluminum and copper films can be described by the conduction through the films and the energy input, or cell heat loss ($\dot{q}$):

$$\rho(\vec{r}) \cdot c(\vec{r}) \cdot \frac{\partial T(\vec{r},t)}{\partial t} = \nabla \cdot [k(\vec{r}) \cdot \nabla T(\vec{r},t)] + \dot{q}(\vec{r}) \tag{3.1}$$

where c is the specific heat capacity, ρ the density, k the coefficient of thermal conduction, and $\dot{q}$ the rate of specific energy input. The spatial variation is expressed by the vector $\vec{r}$. This spatial variation can however be neglected in this case because the aluminum and copper films are assumed to be homogenously coated with the active material (see Section 1.1), and the conduction within the aluminum and copper is assumed isotropic and independent of temperature in the range considered (0 to 60°C). As a result, for a set time interval, Equation 3.1 can be rearranged and reduced to:

$$\Delta T = \frac{q}{c \cdot \rho \cdot \delta} \tag{3.2}$$

where δ is the layer thickness of the copper or aluminum film.

Because the films are assumed to be evenly coated, the specific energy input throughout the cell ($\dot{q}$) resulting from the cell internal resistance is uniform over both the aluminum and copper films ($q_{Cu} = q_{Al} = q$). Thus, the ratio of temperature differences across the aluminum (ΔT_{Al}) and copper (ΔT_{Cu}) can be calculated. A value of unity indicates that a given energy input results in a uniform temperature increase in both layers. The ratio is calculated as a function of the thickness of the copper layer (δ_{Cu}), which is thinner than the aluminum layer.

The ratio of temperature differences in each layer is expressed by Equation 3.3:

$$\frac{\Delta T_{Cu}}{\Delta T_{Al}} = \frac{c_{Al}\,\rho_{Al}\,(1.4\,\delta_{Cu})e_{Cu}}{c_{Cu}\,\rho_{Cu}\,(\delta_{Cu})\,e_{Al}} = 0.999 \tag{3.3}$$

The ratio of the temperature difference in the aluminum and copper foil at a specific energy input is nearly unity, which agrees with the nearly uniform temperature distribution across the jelly roll observed experimentally.

Additionally, the surface temperature of the cell (Figure 3.3) exhibits asymptotic behavior. The general shape of the surface temperature curve is however distorted by local temperature peaks. These local peaks occur as a function of the irreversible heat loss of the cell, which is in turn a function of the internal resistance. The internal resistance, which is a function of cell voltage or state of charge (SOC), varies near the minimum and maximum SOC as shown in Figure 3.5 due to the implemented charging / discharging strategy: constant current, constant voltage.

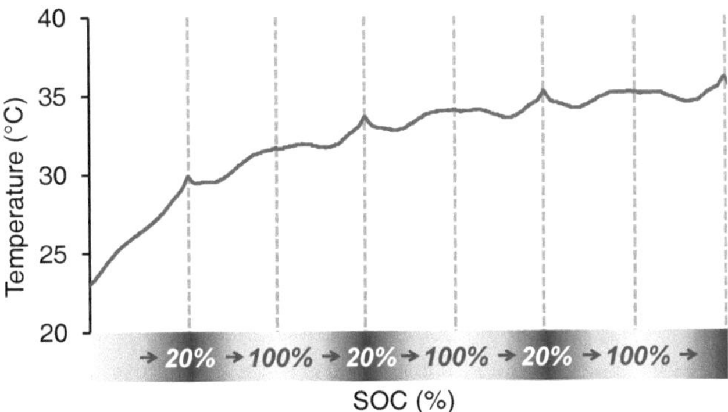

Figure 3.5: Enlarged view of the temperature profile plotted in Figure 3.3. Shown are the temperature spikes at the transition from discharging to charging (20% SOC) and from charging to discharging (100% SOC).

The local temperature peaks indicate inefficient operation near the extremes of battery SOC: for this reason, battery cells are generally not charged/discharged to their limits. A more consistent temperature profile is achieved when the cell is operated between more conservative limits, such as 30 and 90% (or 80%) SOC, which serves to limit local temperature peaks.

The thermal behavior of a prismatic PHEV-2 cell can be described by the first law of thermodynamics.

$$\frac{dU}{dt} = \dot{Q}_{loss} - hA_{\infty}(T - T_{\infty}) \tag{3.4}$$

which states that the internal energy of the cell $\left(\frac{dU}{dt}\right)$ is changed by the heat added and work done on the system. In this case heat is added due to the irreversible heat loss of the cell ($\dot{Q}_{loss}$), and work is done through heat transfer to/from the ambient via Newton's Law of cooling ($hA_{\infty}(T - T_{\infty})$), where h the coefficient of heat transfer to the ambient, and A_{∞} the surface area exposed to the ambient.

Because the measured temperature distribution across the cell is small, a small Biot Number and thus the validity of lumped capacitance is assumed, so that the change in the internal energy in the cell is simplified to $C\frac{dT}{dt}$, where C is the thermal capacity of the cell and $\frac{dT}{dt}$ is the rate of change of the cell average temperature:

$$C\frac{dT}{dt} = \dot{Q}_{loss} - hA_\infty (T - T_\infty) \tag{3.5}$$

Equation 3.5 is fit to the experimental data via Euler's Method by iteratively varying the values for the thermal capacity C and the thermal conductance to/from the ambient hA_∞. The calculated fit is only valid for the set of boundary conditions used experimentally: the time constant varies as a function of ambient temperature (T_∞), initial temperature (T_0) and cell heat loss $(\dot{Q}_{loss})$. When the thermal behavior and more specifically the time constant of the real cell is matched under identical boundary conditions, the SBC then has an identical heat capacity to the actual cell. The first law of thermodynamics is applied in greater detail in Sections 4 through 8, where the module analysis is performed.

3.3. Design and Validation of the Smart Battery Cell (SBC)

The starting point for the design of the SBC is the analysis of the internal geometry. By disassembling the cell, the critical components can be identified. The orientation of internal components (e.g. air pockets, current collector tabs, etc.) influences the temperature distribution. Mimicking the internal cell geometry and individual material properties provides a good starting point for modeling the thermal behavior of the cell. The SBC utilizes the same casing as the real cell in order to provide identical material properties, roughness, and emissivity at the cell surface. As a result, the thermal interface of the SBC to the ambient and other objects is identical to that of the real cell.

Because of the ambiguity of available thermal data for the cell, iterations consisting of constructive design changes and experimental analyses (via the setup presented in Section 3.2) were performed until the SBC matched the thermal behavior of the actual cell.

The final SBC version uses an Aluminum core ("jelly roll") with integrated cooling channels for rapid reconditioning between tests. The heating elements, printed on a thin adhesive pad consisting of eight zones (four per side), is attached on the core. The various zones facilitate the recreation of temperature gradients across the cell if a damaged cell is to be simulated. The heat loss profile that serves as an input for the SBC can reproducibly generated, eliminating the variance caused by individual cells (Section 1.1).

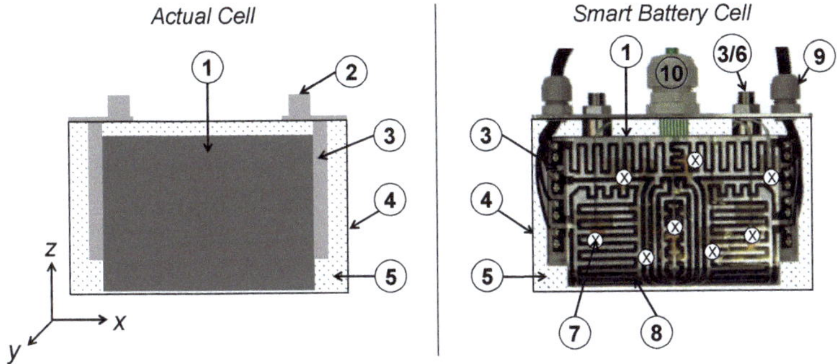

Figure 3.7: Schematic view of the actual cell (left) showing the primary components: the jelly-roll (1), terminals (2), current collector tabs (3), casing (4), and air gaps (5). The SBC (right) shown with the additional features of an internal cooling circuit (6) with connections mimicking the current collector tabs, 16-integrated temperature sensors (eight per side) marked with an "X" (7), 8-heating zones per cell (8) including the required wiring (9), and the connection for the thermocouple wiring (10).

Thin channels are included on the surface of the core under the heating elements to integrate 16 thermocouples (eight per side, Figure 3.7). Because thermocouples rely on a material pairing, obtaining thermocouples from a single material stand helps reduce measurement error [Bernhard 2014]. In cooperation with the selected material distributer, a bulk order was made for the thermocouples used in this research in an attempt to minimize measurement error. All thermocouples were additionally calibrated using a reference power source at the equivalent voltages for five separate temperatures.

Triangular wedges on either side of the core combined with the connections for the internal cooling channels mimic the current collector tabs in the actual cell, while providing a better adhesive surface for the heating elements. The cell lid, containing a flange, slides within the original cell casing and electrical insulation. Care is taken to insulate the data acquisition equipment from the power source in order to avoid electromagnetic interference in the measurements. By integrating the thermocouples in the core, the contact surfaces on the cell exterior are not disturbed, resulting in realistic thermal contact interfaces and therefore realistic measurements of BTMS performance.

The critical design considerations of the SBC are summarized below:

- Robust design and assembly so that the same SBCs can be used for all tests, enhancing the value of the comparison;
- Integration of high quantities of data acquisition equipment within the cell in order to quantify the effect of the thermal management systems on cell aging and capacity loss without affecting thermal contact surfaces;
- Eight individual, over-dimensioned heating zones per SBC in order to simulate damaged or aged cells and extreme heat loss profiles;
- Elimination of electromagnetic interference between switching loads (e.g. heating element wiring) and measurement equipment (e.g. thermocouples);
- An internal cooling circuit, allowing for Hardware in the Loop (HiL) functionality and for rapid reconditioning between tests (multiple trails in rapid succession);
- Utilization of the original cell casing to mimic the thermal contact interface;
- Minimization of the complexity and cost through the use of simple assembly processes and common materials.

Multiple design iterations were undertaken to realize these functionalities [Schneider 2013]. The respective SBC iteration was then tested against the real Lithium-Ion cell using the same experimental setup as presented in Section 3.2. The performance of the final version of the SBC is shown in Figure 3.8.

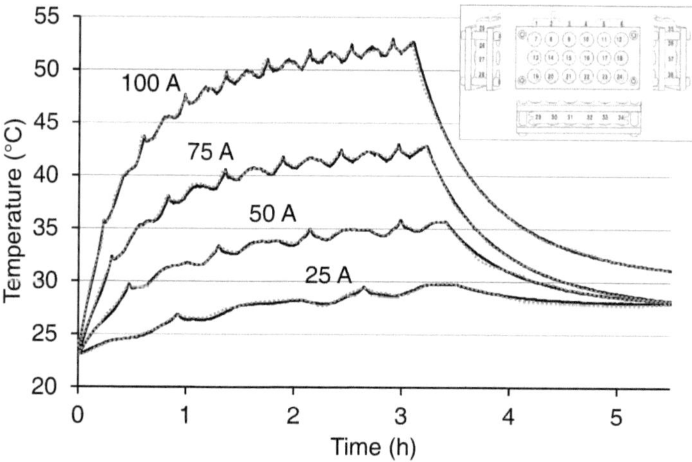

Figure 3.8: Comparison of the average temperature over the surface of a real cell (solid line) and the SBC (dashed line) under the same experimental conditions (Section 3.2). The current input is structured as shown in Figure 3.3.

Using the internal geometry of the cell as a reference, and accurate hardware-based model of a PHEV-2 cell is realizable. The accurate dynamic thermal performance of the SBC indicates an equivalent heat capacity to the real cell. Using the integrated thermocouples of the SBC, the effect of various thermal management concepts on the cell can be quantified experimentally. By means of the internal coolant circuit of the SBC, reproducible initial conditions are quickly reached, allowing for far more trials to be performed in a similar time.

As discussed in Section 1.1, many manufacturers utilize so-called battery modules as building blocks for a complete battery system. These building blocks can then be combined in various configurations to realize an array of battery systems for different vehicle. Because the module is the building block of many battery system variations, the module-level is the most universal level at which to compare thermal management concepts. To do so, eight SBC are combined to a module as described in Section 4.

3.4 Novelty of the Developed Method

The SBC presents a spatial measurement resolution not available in the literature. The undisturbed thermal contact surface and integrated measurement equipment facilitate reproducibility that is difficult to achieve with other temperature measurement techniques and real Lithium-Ion cells.

By eliminating the active chemistry, thermal management concepts can be compared more efficiently under extreme conditions with minimal safety risks. The undisturbed thermal contact surface and integrated measurement equipment facilitate reproducibility that is not always achievable with other temperature measurement techniques and real Lithium-Ion cell. Using the internal cooling circuit, the reconditioning time is drastically reduced.

These benefits and those discussed in Section 3 make the SBC a novel alternative to real Lithium-Ion cells.

Analysis

4. Reference Module for the Experimental Comparison

The SBC (presented in Section 3) models an *individual* cell. The analysis of thermal management concepts presented in this work is however not conducted at cell-level, but rather at module level (Figure 1.2). The analysis at module-level allows for the temperature homogeneity between cells to be compared, providing information regarding battery system aging not possible to determine at cell-level (see Section 2.1). Results from module-level analyses can be scaled to various battery system sizes. To facilitate the analysis at module-level, a universal basis-geometry is adapted [Smith 2014]. This basis-module consisting of eight SBCs is combined with a battery thermal management system (BTMS) to form a so-called "technology demonstrator." The use of a basis-geometry and a custom climate chamber guarantees reproducible boundary conditions, meaning that only the effect of the BTMS on the cells is compared across various technology demonstrators.

The following section presents the basis-geometry and experimental setup used. Thereafter, the reference system—a module without a BTMS—is presented and analyzed. This reference system is the basis for the analyses of the various BTMS technology demonstrators (Sections 5 through 8).

4.1. Experimental Setup and Procedure

To accurately compare thermal management systems, the only variable should be the systems themselves. Thus, a basis-module created in the BMBF research project *eProduction* was adapted [Smith 2014]. The battery module was conceived to publically and abstractly showcase various technologies relevant to the production of high-voltage battery systems for electric and hybrid-electric vehicles; therefore, the layout is relatively simple.

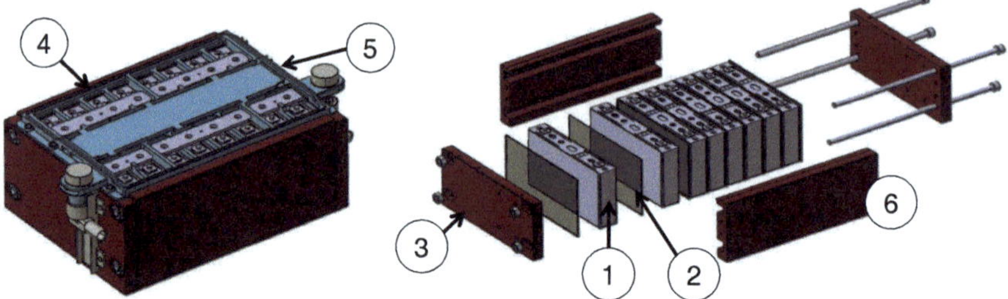

Figure 4.1: the basis-module (left) as well as an exploded view (right). Shown are the cells (1), the spacers (2), the compressive end plates and threaded bolts (3), the electrical contacts or bus-bars (4), the so-called bus-bar carrier (5) and the side plates (6) [Smith 2014].

The basis-module consists of eight prismatic cells, which are replaced with SBCs in the case of this research. In place of the electrical contacts required for actual cells are the data acquisition, power supply, and hydraulic connections for the SBC. The casings of actual battery cells must be electrically insulated to one another to prevent short circuiting. For this research, the 3 mm polycarbonate spacers used for transport packaging by the cell manufacturer were implemented to simulate the required electrical insulation. These spacers are used for the reference system (no thermal management, Section 4.2) and the cooling plate experiments (Section 5). In the additional experiments, these spacers are replaced with the active spacers under study.

Actual prismatic cells produce large expansive forces during operation that change with the state of charge (SOC) and over the lifetime of the cell; therefore, compressive plates are used to counter this effect. The SBC does not generate a mechanical force, but the resulting force is mimicked by means of a high compressive force between over-dimensioned end plates.

This basis-module is adaptable enough to accommodate a multitude of thermal management concepts, yet eliminates superfluous variables in the experimental trials, allowing for a comparison of just the battery thermal management systems (BTMS). Various BTMS concepts are integrated with the basis-module to form a so-called technology demonstrator. Various technology demonstrators are utilized throughout this work to compare thermal management concepts. The technology demonstrators reflect an early development stage and conceptual analysis: they are not optimized, but rather analyzed based on the evaluation criteria presented in Section 2. Their primary function is to identify promising concepts for future analyses.

Ambient boundary conditions must also be consistent and reproducible in order to eliminate additional variables and isolate the effect of the BTMS on the cells. Therefore, all experimental technology demonstrators are placed on a 25 mm Polyoxymethylene (POM) base, and all other surfaces are left exposed. The primary reason for this decision is to avoid errors caused by insulating experimental setups that are very distinct in their layout: the same thickness and coverage over the entire demonstrator cannot be guaranteed, resulting in varying heat fluxes. On the other hand, the boundary condition of natural convection chosen must be guaranteed by the climate chamber, and convection and edge effects on the technology demonstrator will be observed in the experiments. However, an exposed setup facilitates rapid conditioning to the initial conditions for the experimental trials. Furthermore, safety is increased by giving the operator visual access to the setup. A schematic of the thermodynamic boundary conditions of the setup is shown in Figure 4.2. The custom-designed climate chamber is described in detail in Appendix C.

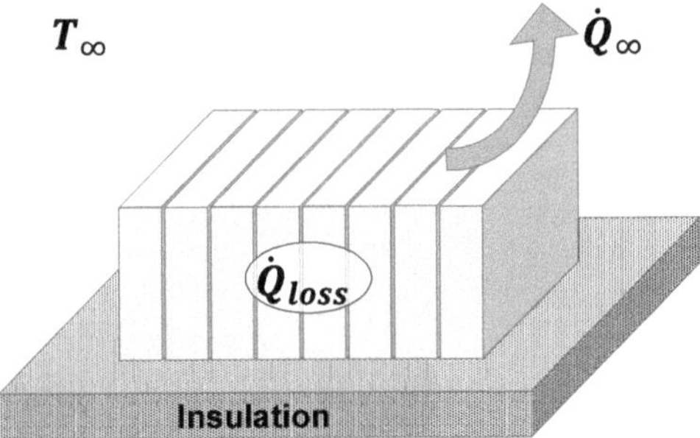

Figure 4.2: A schematic of the eight cell reference module on the insulated base. Shown are the thermodynamic boundary conditions including the heat loss of the cells ($\dot{Q}_{loss}$), the heat transfer to/from ambient ($\dot{Q}_\infty$), and the ambient temperature (T_∞), which is held constant throughout the trial.

The Aluminum end plates of the basis-module (20 mm thick) are compressed with four M8 threaded bolts and nuts. The bolt head and nut are tightened to 9 Nm for each experiment. This tension is selected because it replicates the tension between cells, and is at the upper limit for the bolts. By using a constant tension, the contact surface between cells is intentionally eliminated as a variable in this research. The SBCs are connected (thermocouples, power source and hydraulic lines) in their corresponding positions. The climate chamber temperature is set and, in combination with the hydraulic circuit in the SBC, the complete technology demonstrator and ambient are climatized to the desired test temperature. When the measured temperatures of each thermocouple are within 1°C of one another, the trial is manually started.

A heat loss profile for the SBCs is additionally required as an experimental input. The heat loss profile for the SBC represents the irreversible loss (internal resistance) and is calculated as a function of the instantaneous current, voltage, and state of charge of the test vehicle's battery system. Internal resistance is also a function of temperature, so for this model, the initial temperature is set equal to the ambient temperature. The temperature input to the cell model (the ambient temperature), is held constant throughout the test instead of coupling the instantaneous cell temperature to the cell model. Therefore, the energy input is constant between all trials with various technology demonstrators. As a consequence, the exact reaction of the cell to the temperature change is not depicted. The discrepancy between the actual cell heat loss and calculated cell heat loss is minimal in the trials at above 20°C ambient. As the ambient temperature approaches 0°C, the non-linear increase in cell internal resistance results in a larger discrepancy. This discrepancy is however accepted, because using the current set-up, the *exact* energy input is known and is *constant* between trials as are the other boundary conditions.

In order to eliminate confusion stemming from the non-linear relationship between ambient temperature and thermal energy input to the cells, the three ambient conditions are instead referred to as Case A ($T_\infty = 0°C$), Case B ($T_\infty = 20°C$), and Case C ($T_\infty = 40°C$). The heat loss profile ($\dot{Q}_{loss}$) and the resulting cumulative energy input (Q_c) used for Cases A through C are shown in Figure 4.3.

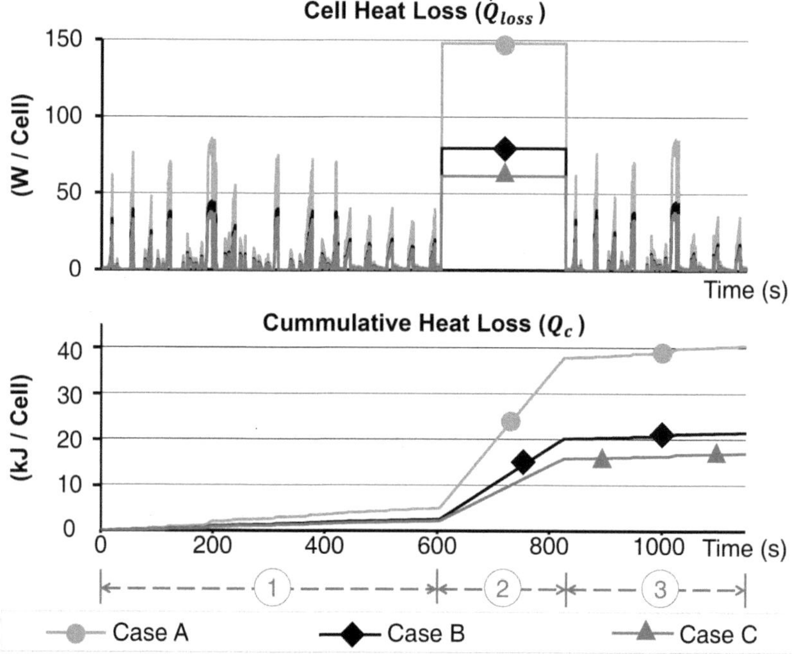

Figure 4.3: The heat loss profile used in the experimental trials (top, $\dot{Q}_{loss}$) and the resulting cumulative energy input (bottom, Q_c) for the three Cases considered. The same x-axis (time) is used for both plots. The three segments of the heat loss profile designed to explore various characteristics are: aggressive city and country driving section (1), rapid charging rate / aggressive racetrack performance (2), and an additional aggressive city and country drive (3).

The same heat loss profile ($\dot{Q}_{loss}$) shown in Figure 4.3. is used for each of the eight SBCs, so that a spatially uniform heat input can be assumed.

The heat loss profile used in all experiments stems mostly from actual vehicle trials, and is designed to explore the dynamic behavior of the thermal management system. The profile begins with an aggressive city and country driving section (1), followed by simulated rapid charging or aggressive racetrack performance (2), and an additional aggressive city and country drive (3). Both the first (1) and third (3) portions of the drive cycle are from actual vehicles trials with a luxury, high performance PHEV. The second portion (2) of the drive cycle is calculated to represent the worst case. The driving cycle represents a demanding yet feasible drive cycle for a high-performance PHEV.

The climate chamber temperatures for each case is maintained throughout the duration of the trial. Relative humidity is measured but not varied. A new trial is started when the aforementioned start criterion is reached (all thermocouples within 1°C). Three trials are performed at each set of boundary conditions. The trials at a given ambient temperature are performed sequentially (i.e. three times at 20°C, then three times at 40°C) in order to increase experimental efficiency by not reconditioning the entire chamber. The hydraulic circuit of the SBC also serves to cut reconditioning times significantly by quickly reconditioning the SBC-module. The temperature data from the climate chamber and the SBC are logged every half-second. This experimental procedure is repeated for the analyses of all technology demonstrators, with the addition of the parameters coolant flow rate and coolant inlet temperature for the BTMS concepts.

4.2. Analysis of the Reference Module

Particularly from the standpoint of vehicle integration, producibility, and economic viability, an active thermal management system can only be justified if the thermal behavior of the cells improves. Without a significant improvement in vehicle range and/or system lifetime, the increased production complexity and cost are unwarranted. Thus, the reference case for this research is an eight-SBC basis-module without a thermal management system. Because all boundary conditions are maintained for all other trials, any difference observed is due to the BTMS alone.

For the three cases tested, the energy input into the module is different. The different energy inputs result in a different temperature increase over the module, as shown in Figure 4.4.

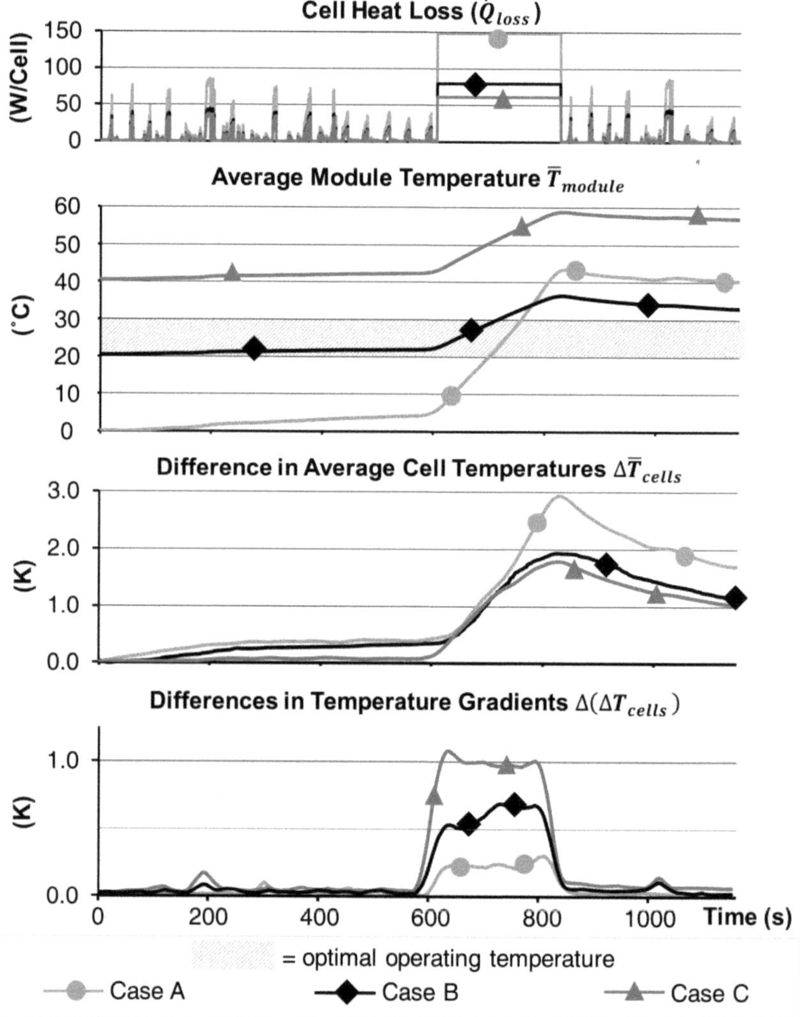

Figure 4.4: The cell het loss profile (input) for cases A through C is shown at top, followed by the three criteria for thermal performance (Section 2.1): the average module temperature $\bar{T}_{module}$, the temperature difference across the module $\Delta\bar{T}_{cells}$, and the differences in the magnitude of the cell temperature gradients $\Delta(\Delta T_{cells})$. All plots share the same x-axis and are shown for cases A through C for the reference module.

In the case of 0°C ambient temperature, the impact of the selected cell model is shown: in the current set-up, the instantaneous heat loss is not a function of the instantaneous cell temperature, but rather of the initial/ambient temperature. The internal resistance of a real Lithium-Ion cell would decrease as the cell temperature increased; therefore, the temperature increase shown in Figure 4.4 is higher than if a dynamic cell model utilizing the instantaneous temperature were implemented. However, the advantage of the cell model used is that for each trial at a certain ambient temperature—regardless of the BMTS used—the exact same energy input results, allowing the BTMS to be compared directly based on energy transfer. Additionally, the variations in cell heat loss as a function of ambient temperature are much more significant near 0°C: between 20 and 40°C ambient temperature, the cell internal resistance and therefore the heat loss does not fluctuate as significantly, meaning that the choice of cell model is less significant in this range.

The average temperature of the module as a function of time (thermal evaluation criteria, Section 2.1) is used to model the behavior of the module based on the first law of thermodynamics. The first law of thermodynamics is applied to approximate the magnitude of the heat transfer to/from ambient of the module-level reference system assuming the module as an incompressible solid:

$$C\frac{dT}{dt} = \dot{Q}_{loss} - hA_\infty(\bar{T}_{module} - T_\infty) \tag{4.1}$$

where $\frac{dT}{dt}$ is the rate of change of the module average temperature, $\dot{Q}_{loss}$ the irreversible cell heat loss, h the coefficient of heat transfer to the ambient, and A_∞ the surface area exposed to the ambient. The quantity hA_∞ is the thermal conductance, which is the inverse of the thermal resistivity to the ambient (R_∞). The solution is approximated by fitting the unknown coefficients using Euler's Method to the experimentally measured average module temperature. To do so, the thermal conductance (hA_∞) and the heat capacity (C) are iteratively varied until the difference of the squares of the experimentally measured average module temperature $\bar{T}_{module}$ and the calculated solution are minimized. The resulting goodness of fit between experiment and calculation is expressed by the coefficient of determination (r-squared value).

The initial value of heat capacity for the iterative fit process is approximated based on the materials in the SBC, and the initial value for the thermal conductance based on the surface area to ambient and the natural convection boundary condition to the ambient. The parameters determined from the fitting process are shown in Table 4.1. The heat capacity is rounded to the nearest hundred, as further accuracy assumed unrealistic. Additionally, the heat capacity is assumed to be constant over the temperature range considered (0 to 60°C), because within this range, the specific heat capacity of aluminum (the primary component by volume and mass), varies within the error tolerance of the fitting process. The spread of the thermal conductance values to/from the ambient are also assumed similar enough within the bounds of the error of the fit process. Despite the uncertainty, an adequate fit is achieved (r-squared value near 1).

Table 4.1: Iteratively calculated values for the heat capacity (C) and thermal conductance to/from ambient (hA_∞) at the three cases analyzed for the reference module. The goodness of fit of the numerical solution is also shown (r^2).

Case	C (J^{1}K^{-1})	hA_∞ (W^{1}K^{-1})	Fit (r^2)
A	6500	3.0	0.999
B	6500	3.4	0.998
C	6500	3.5	0.999

The results in the table indicate nearly constant coefficients C and $(hA)_\infty$, which agrees with the assumption of constant heat capacity over the temperature range considered, and a constant coefficient of heat transfer (h_∞) because the area exposed to the ambient is constant between cases. The fitted solution is most sensitive to the value of the thermal conductance in the regions where the instantaneous cell heat loss is low. The fitted solution is not plotted in Figure 4.4 because the difference is not apparent.

Using the fitted parameters, the three terms in Equation 4.1 can be separated and the effect of each term shown independently (Figure 4.5).

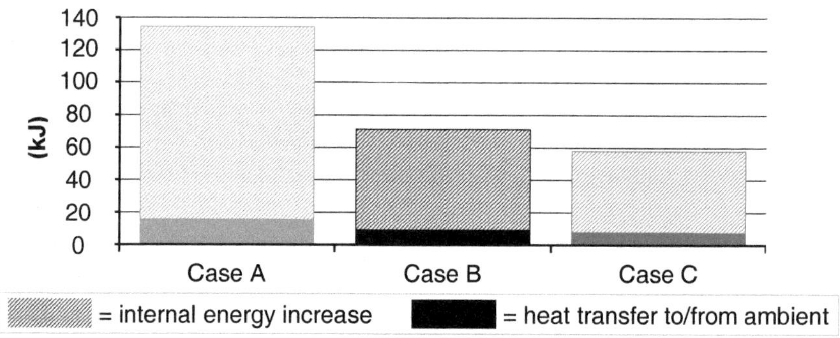

Figure 4.5: Contribution of the internal energy increase in the module and the heat transfer to/from the ambient based on the total energy input (total column). The average value over the experimental time is shown.

Figure 4.5 shows the relationship between the total energy input and the internal energy increase as well as the heat transfer to/from ambient. The determination of the coefficients thermal conductance and heat capacity allow for a comparison of the thermal conductance to/from the ambient and/or a BTMS over diverse systems tested at different boundary conditions

The fit is relatively good because of the small spatial temperature distributions over the module, which indicates that a lumped mass assumption is acceptable. The spatial temperature variations are greatest at high heat losses; thus, the fitted solution of the average module temperature is most accurate in "normal" operating conditions (10 to 30 W per cell heat loss). The spatial temperature distribution is characterized separately by $\Delta\bar{T}_{cells}$ and $\Delta(\Delta T_{cells})$. The necessity of both criteria can be seen: the two temperature distribution criteria react differently based on the cell heat loss. In the case of the reference module, however, the magnitude of $\Delta(\Delta T_{cells})$ is insignificant compared to $\Delta \bar{T}_{cells}$. Additionally, as presented in Section 3, the thermocouples carry a measurement uncertainty of $+ 1.5$ K, meaning the measured $\Delta(\Delta T_{cells})$ of around 1 K can be neglected for this reason as well. The reference module with no thermal management is susceptible to inhomogeneous aging primarily at high heat loads, where the average temperatures of the cells are up to 3°C apart. The exact degree of aging due to a 3°C temperature difference can only be quantified if the use behavior is know: if the 3°C difference were assumed to be continuously applicable (24 hours per day), the relationship between capacity loss and temperature is described by the Arrhenius equation (Section 1.1). For practical purposes, a 3°C difference at high heat loads is acceptable.

The module without thermal management is the reference system for the analysis of various BTMS concepts: thermal management concepts that do not sufficiently improve the thermal state of the cells over the reference case based on the three criteria for thermal performance are superfluous. In Sections 5 through 7, variations of three different thermal management concepts are compared to one another across all use cases. Thereafter, the most promising thermal management concept variations from each Section (5 through 7) are compared to one another and the reference module.

5.   Investigation of Various Cooling Plate Designs

Though many diverse cooling plate layouts are possible, some common features and design guidelines exist. A cooling plate utilizes a coolant to transfer heat to and from the battery. Currently, a water-glycol mixture or a refrigerant (such as R134a or R1234yf) are commonly used in production vehicles. In either case, the leak-tightness of the cooling plate (and complete circuit) is critical to protect against coolant leaking on the battery and potentially causing a short circuit. Because absolute leak-tightness does not exist, an acceptable value and correspondingly an appropriate test method must be selected [Kritzer 2013]. Leak-tightness is influenced by cooling plate layout and production technique selection to ensure that, for example, the joining of two pieces meets the leakage specification. Forces within the cooling plate are a function of the pressure created by the coolant in the cooling plate channels, which is again a function of the channel size and the required inlet pressure to overcome the pressure drop from pump outlet to inlet. Selection of the appropriate production technique and the corresponding cooling plate design to meet the specified leak-tightness is critical.

A cooling plate must also thermally contact the cells. Better thermal contact translates directly to more efficient BTMS performance [Fletcher 1988]. This contact interface is however complicated by the required electrical insulation between cells, and by the manufacturing tolerances of the battery cell casings and of the cooling plate. Thus, the acceptable deformation during the production process and the resulting flatness tolerance for the final product must be established. As with any production process, tighter tolerances generally correlate to additional production steps and higher production costs. The tolerance compensation and electrical insulation of the cells occur at the same interface: between cooling plate and cell. This additional layer(s) creates an additional contact resistance. Thus, assuming the contact area is given, either the thermal conductivity of the material must be raised and/or the thickness of the layer must be decreased.

In all cases, air gaps should be avoided, as they can create local hot-spots and temperature gradients across the module. Very thin (0.01 to 0.1 mm) electrically insulating foils or coatings from the consumer electronics industry generally have low thermal conductivity (0.1 to 0.2 $W^1m^{-1}K^{-1}$), but are not ductile and therefore cannot compensate for short-wave tolerances. Thicker thermal pads have conductivities over an order of magnitude higher (3 to 5 $W^1m^{-1}K^{-1}$), and are available in various thicknesses to compensate tolerances; however, the force required to compress the pad does not scale linearly with surface area. Therefore, the module and cooling plate must be designed to handle the resulting forces. Thermal pastes with similar conductivities to the pads are also prevalent due to the lower costs versus a pad and the lower forces required to compress an initially liquid material. The electrical resistance is however not guaranteed by the paste because of the dispersion or uneven spreading during application. Therefore, in order to secure against a short circuit, an electrically insulating film must be used in conjunction. Thermal pastes also pose a challenge in the production and most likely require automation to achieve adequate process stability. In all cases, a reduction of the layer thickness and reduction in the air gaps can additionally be achieved through enhanced contact pressure between the cells and the cooling plate [Daubitzer 2012, Weileder 2012, Hirsch 2013]; however, as mentioned, the module and cooling plate must then be designed for the resulting mechanical load.

Guided by the basic requirements of leak-tightness and the electrical insulation of the cells, various cooling plate layouts and possible production techniques are discussed in Section 6.1. The cooling plate technology demonstrators are experimentally analyzed in Section 6.2. The analysis is discussed, and the lessons learned and recommendations for the cooling plate for a high-performance PHEV are made in Section 6.3.

5.1. Layout and Considered Production Techniques

All cooling plate designs presented in the feasibility study use a 50/50 water-glycerol mix common in the automotive industry. Various production techniques and cooling plate layouts were assembled and analyzed to meet the criteria of leak-tightness: no visible leakage at an internal pressure of 2.5 bar, and under 0.1mm of deformation over the contact surface. In normal operating conditions, much lower internal pressures are expected, but this safety factor protects against pressure surges. All cooling plates were manufactured by project partners from the research project *eProduction* [BMBF 2014]. were therefore tested for leak-tightness with air using a mass-flow measurement machine setup according to the manufacturer's directions. The cooling plate was additionally placed in a water bath to visualize leakage. Four joining techniques and variations thereof were investigated and evaluated, as shown in Table 5.1.

Friction Stir Welding and Variations Thereof

Friction Stir Welding (FSW), normed under DIN EN ISO 25239, utilizes a rotating tool to "stirs" the materials together below the material melting points and create a solid state weld [Guo 2015, Khandkar 2003]. No filler material or protective gasses are required. Because the welding occurs below the material's melting point, lower thermal deformation and a higher quality weld generally result [Lohwasser 2009, Tang 1998]. The rotating torque combined with a high mechanical force (3.5 kN in this analysis) requires significant bracing of the components being joined [Reisgen 2012]. This type of welding has been implemented in ship building, fuel tanks and other high strength applications [Lohwasser 2009]. Compared to conventional fusion welding techniques, FSW is more universally applicable and economical to realize [Guo 2015, Mishra 2007]. FSW can join materials of different types [Nelson 2000] or different alloys [Nagano 2001]. The process can be highly automated, but the upfront investment is relatively high due to the need for custom made equipment for the tooling and bracing, making this process more viable for larger production volumes [Lohwasser 2009].

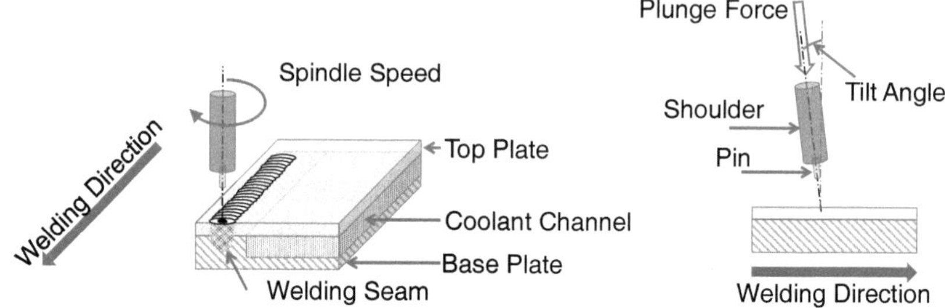

Figure 5.1: Schematic of the basic components and parameters of friction-stir welding (FSW). Adapted from [Lohwasser 2009] and [Eireiner 2006].

In addition to the presented conventional FSW, two further variations were analyzed: Delta-N and BondWELD (adhesive-fixed). Delta-N differs from standard FSW in that only the tool pin (and not the shoulder) rotates; therefore, Delta-N generally results in a lower thermal input. However, the mechanical compression was increased by 1 kN to ensure a leak-tight weld seam. As a result, no difference in deformation could be measured on the cooling plate geometries tested. BondWELD uses adhesive strips to attach the components to be bonded, meaning that no additional mechanical bracing is needed between the components [Reisgen 2012].

In all cases analyzed, the welding parameters play a large role in the deformation and leak-tightness: for example, a higher tool rotational speed combined with a higher linear tool speed served to reduce deformation by over 20%. The possible cooling plate geometries and weld seam paths are a function of pin diameter and machine layout (number of axes). During the trials, the length of the weld seem drastically influenced deformation because of the large thermal input, especially considering the small size of the cooling plate. As a result, especially the corners of the rectangular plate deformed. Start and end points of the weld (i.e. where the pin must sub- and emerge from the weld seem) were most critical for leak-tightness. The use of cold rolling after the welding of the plates resulted in acceptable deformation and surface finish. FSW and variations thereof are most suited for a cooling plate that must maintain leak-tightness under high mechanical loads. Adequate flatness at the contact surface to the module was achieved using thicker cooling plates or cold rolling.

Electron Beam Welding

Electron-beam welding (EBW) uses a cathode ray of high density and current under vacuum to accelerate electrons to 50 to 80% of the speed of light, and form a beam via an electromagnetic field [Schubert 2010, Schultz 1993]. The thin, highly concentrated beam induces thermal conduction within the work piece, instead of just on the surface, and can be reproducibly created [Schultz 1993]. The use of a laser provides a very concentrated, deeply penetrating and narrow weld seem that affects a much smaller area than FSW, theoretically resulting in less deformation [Lohwasser 2009]. As with

FSW, no fillers or protective gasses are required. In this research, only like-material pairings were considered, but EBW has been used for dissimilar materials [Sun 1996].

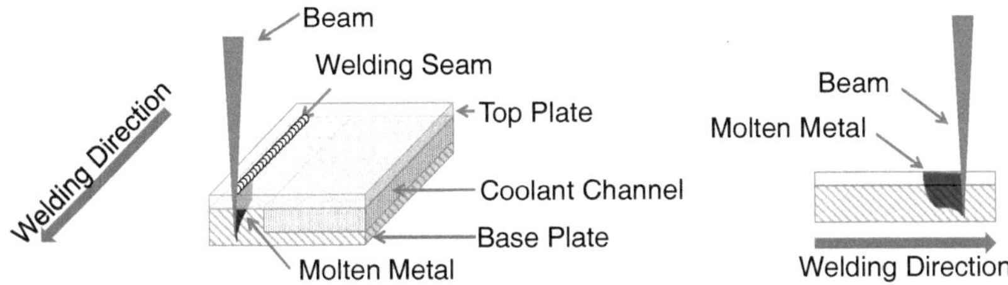

Figure 5.2: Schematic of the basic components of the weld seam for Electron Beam Welding (EBW). Adapted from [Schubert 2010].

As for FSW, the start and end of the weld seam exposed potential leakages in the analyses performed, requiring the weld-path to be adapted. The linear welding speed in the trials was generally higher than for FSW. Even though no mechanical pressure is applied, the thermal input was still significant enough to cause deformation in the thin cooling plates analyzed comparable to those joined with FSW. As with FSW, additional flatting by cold rolling yielded adequate flatness.

Adhesion

Adhesion relies primarily on the von der Waals and electrostatic forces between two materials [Bhomik 2015]. In contrast to the solid state welding techniques considered, a high thermal input into the material can be avoided in many cases. Adhesives are widely used in many industries to join parts of various materials while saving weight [Bhomik 2015]. Good adhesion depends on the material pairing and surface preparation, the application of the adhesive, and the bonding or curing process [Ebnesajjad 2011]. Adhesives are not suited to materials that undergo large volumetric changes during use (i.e. PA when taking up moisture). Leak tightness and corrosion resistance can be improved with a continuous slot through the contact surface, which serves as a reservoir for the adhesive during joining, and lengths the path for impurities to penetrate the joint [Bhomik 2015].

With proper surface treatment and curing time, the cooling plates tested performed well in leak-tightness tests. The process stability of the surface treatment must however be guaranteed, as adhesive fracturing was noticed at very low loads when surface treatment was neglected. For the technology demonstrators, cooling plate parts were first cleaned with Methyl-Ethyl-Ketone (MEK), then sandblasted with F100 size grains (FEPA Standard), and again cleaned with MEK. A two-component adhesive, Araldite 4415, was selected because of its suitability for the desired application [BMBF 2014].

In general, adhesives and the surface treatments emit hazardous vapors: MEK, for example, is an irritant and flammable substance. Accommodations must therefore be made in a production in order to protect workers and meet environmental standards. Additionally, the curing time and/or temperature must be accounted for in the production and logistics. In this way, a cooling plate could possibly cure while waiting to be assembled into the battery system.

Table 5.1: Comparison of various production techniques. Adapted from [BMBF 2014].

Specification		Friction Stir Welding (FSW)			Adhesives	Electron Beam	Soldering
		Conventional	Delta-N	BondWeld			
Loading	Normal to sealing surface	++	++	++	*function of contact surface*	+	*function of contact surface*
	Along sealing surface (shear)	++	++	++		+	
Design Flexibility		o / -	o / -	o / -	++	o	o
Material	Possible materials	metals	metals	metals	all	metals	metals
	Material combinations	+	+	+	++	- / o	o
	Flexibility in material thickness	o	o	o	++	o	-
	Flexibility in material tolerances	++	++	++	o / -	+	++
Corrosion Resistance		+ / o	+ / o	+ / o	+ / o	+ / o	++
Dimensional Stability during Process		- -	- -	- -	+/++	o	o / -
Estimated Lifetime		++	++	++	o	o / +	+
Pretreatment Process		o	o	o	+	o	+
Investment		- / - -	- / - -	- / - -	o	o / -	-

Multiple iterations to optimize the parameters of the aforementioned joining techniques were conducted in cooperation with partners in the research project *eProduction*, with the goal to maximize leak-tightness and minimize deformation through the joining process. The process of soldering itself was not analyzed in detail within the scope of this project, but sample cooling plates were tested experimentally for thermal performance and energy consumption.

Cooling Plate Technology Demonstrator 1:

The first cooling plate technology demonstrator utilizes circular fluid displacing bodies (FDB) within the flow channel in order to provide mechanical stability to the cooling plate, provide a joining surface for the top plate, and to locally increase the velocity of the coolant. By reducing the channel cross section with a FDB, the local flow velocity and therefore Reynolds Number are increased. As a result, the local heat transfer to the channel walls (in the z-direction, Figure 5.3) is increased. Simultaneously, the heat transfer along the surface of the circular FDB is altered: the local heat transfer is increased at the flow stagnation point, decreases along the laminar boundary layer to a minimum at the separation point ($\theta = 80°$ in laminar flow), and increases again due to mixing in the wake [Cengal 2007].

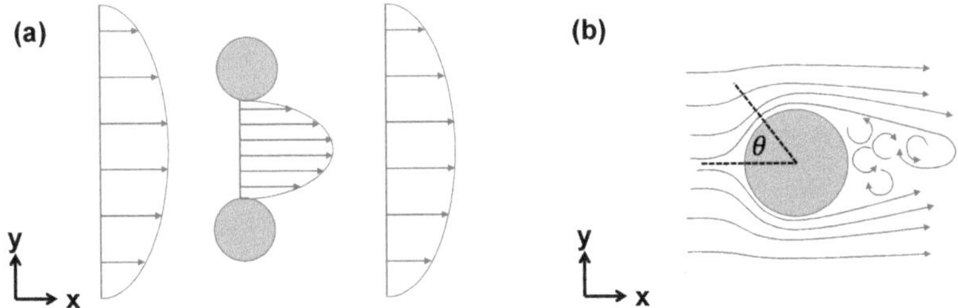

Figure 5.3: The increased flow velocity between two FDBs (a) resulting in the enhanced heat transfer to the FDB-walls, and the flow pattern around a single FDB (b) showing the angle from the stagnation point (θ).

The FDB further serve the cooling plate design by providing a contact surface for various production techniques, thus simultaneously providing the mechanical structure to resist internal pressure spikes and external mechanical forces. The actual technology demonstrator consists of two main parts: a 4 mm thick base-plate milled with 3 mm deep channels joined to a 1 mm thick top-plate. EN AW-5754 (H22) aluminum was chosen for the feasibility study due to its weldability and structural properties. The milled base-plate is meant to simulate a casted, molded, deep-drawn, or hydro-formed part if the production volume was sufficiently high. Depending on the joining technique used, a polymer could be used for the base-plate, and an aluminum top-plate could be attached.

During preliminary tests, the cooling plate was joined via adhesion, FSW, BondWELD, and EBW. The FSW and BondWELD, due to the tool pin, caused leakages sporadically when attempting to contact the top-plate to the FBDs. EBW, when using an optimized path and parameters, proved leak-tight within the specification. For this cooling plate, a method was developed at the Welding and Joining Institute (ISF) of the RWTH Aachen to also contact the inlet and the outlet using EBW and in the same bracing. Adhesion, using a two-component adhesive on an etched and cleaned surface, also proved robust and exceeded the leak-tightness specification. The inlet and outlet were also joined with the same adhesive and surface preparation in this case. The position of the inlet and outlet, shown in Figure 5.3, create a flow pattern known as an "I-flow."

Because of the identical layout and materials (besides at the welding seem or adhesive joint), only the plate joined via EBW was experimentally tested.

Cooling Plate Technology Demonstrator 2:

Another technology demonstrator was created to test a different layout: two symmetrical Aluminum half shells (also EN AW-5754 aluminum). The theory behind this design is that increased amount of similar parts (i.e. two half-shells per cooling plate) will increase part output and make processes like casting, molding, deep-drawing, or hydro-forming more economically viable. Using two like-material pieces means more joining techniques can be implemented. The thicker pieces (2.5 mm), as opposed to the 1 mm top-plate in cooling plate technology demonstrator 1, are more resistant to thermal deformation during the welding processes. Additionally, the narrow channel of constant width meanders multiple times under the cells (as opposed to cooling plate technology demonstrator 3). The design leaves more solid material, allowing for greater mechanical strength and eventual structural integration with the vehicle.

This cooling plate was also joined via adhesion, FSW, BondWELD, and EBW in preliminary tests. The largest deformation was realized using FSW and BondWELD, but with the addition of a further rolling process, an acceptable flatness was achieved with simultaneous de-burring. Though additional steps are required for the use of this technique, the high strength of the FSW weld makes such a cooling plate attractive if structural loads will be transferred through the cooling plate, for example as part of the vehicle structure. EBW again met the specification with less deformation than FSW, implying that an optimization of the welding parameters could eliminate the need for a rolling process. The same adhesive and surface preparation as for technology demonstrator 1 were used, and again the leak-tightness specification was exceeded. The adhesion process also causes no deformation, but often requires long curing times. An optimized production, however, may be able to minimize the impact of the curing process.

The inlet and outlets are attached on the same side (so-called U-flow) of the cooling plate, as shown in Figure 5.3, resulting in a so-called "U-flow" pattern. Because of the identical layout and materials (besides at the welding seem or adhesive joint), only the plate joined via adhesion was experimentally tested.

Cooling Plate Technology Demonstrator 3:

The final design was chosen to represent the state of the art in cooling plate manufacturing. Soldering is widely implemented for heat exchanger production in the automotive industry. By using rolled or extruded tubes, very thin, light-weight cooling plates can be realized. This technology demonstrator consists of multiple aluminum pieces including two extruded aluminum profiles, two manifolds, and two inlets soldered together in one step. The production of such plates was not analyzed within the scope of the research project *eProduction*, but a sample was obtained for testing.

The extruded aluminum profiles only partially cover the base of the module, as shown in Figure 5.3. The "partial" contact to the battery module has been chosen to determine if the weight savings are realized at the cost of thermal performance. A U-flow inlet/outlet position is used.

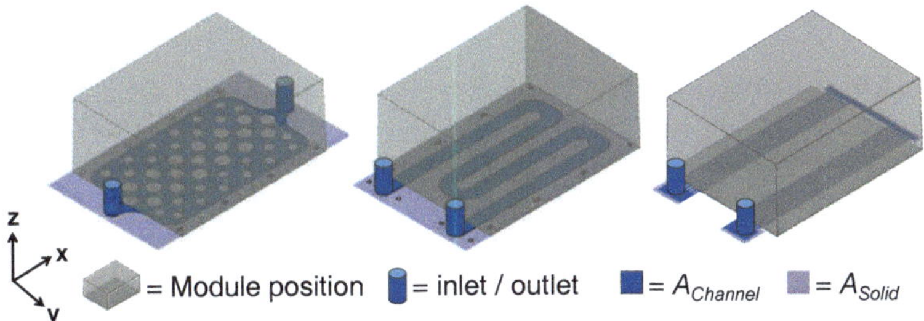

Figure 5.4: The three cooling plate technology demonstrators (1 through 3, from left to right) analyzed experimentally. The Cartesian coordinate directions are shown as a reference.

For the experimental trials, a Berquist Gap Pad 5000 S35 provides the thermal contact and electrical insulation between the module and cooling plate. The compression force is held constant at all four corners of the module for each trial, and the cooling plate is stiffened by the setup presented in Figure 5.4.

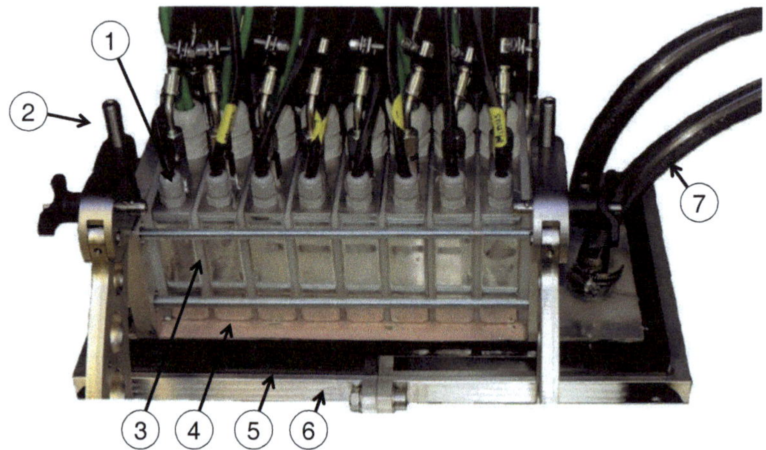

Figure 5.5: Experimental setup for the experimental analysis of the cooling plate technology demonstrators. Shown are the module (1), the compressing-arms for applying the contact force to the cooling plate (2), the spacers (3), the thermal gap pad (4), the thermally insulating base (5), the stiffening-plate (6), and the coolant connections (7).

Each cooling plate technology demonstrator is fixed in a thermally insulating base and backing plate, as for the trials using the reference case (Section 4). The backing plate provides stiffness against the compression force, eliminating cooling plate stiffness as an experimental variable. The coolant inlet and outlet hose-lengths between the measurement point and cooling plate are also held constant. The results and analysis thereof are presented in the next section.

5.2. Multi-Factorial Analysis of Three Cooling Plates

The three cooling plate designs presented in Section 5.1 are experimentally analyzed in this section. The evaluation criteria presented in Section 2 are utilized, and the basis for the comparison is the reference module without a BTMS (Section 4). The thermal performance and theoretical pump power are experimentally measured at various ambient temperatures, coolant flow rates, and coolant inlet temperatures using a 50/50 water-glycerol. In all cases, the boundary conditions are held constant throughout the trial. To facilitate clarity in the plots throughout the research, the case of 20°C coolant inlet temperature and 5 L^1min^{-1} flow rate is shown. As expected, a lower coolant inlet temperature and higher flow rate decreases module average temperature. This trend is consistent over all trials. The multifactorial analysis is presented below.

Thermal Performance

Figures 5.6 through 5.8 show the three evaluation criteria for thermal performance as well as the heat loss profile of each cell in the module at the three cases analyzed. The cooling plate brings the average module temperature nearer to the optimal temperature than the reference case without thermal management; however, the spatial temperature variations created are generally larger than without thermal management. Such behavior is expected: the insulated base of the reference case is swapped for a cooling plate with water-glycerol. Additionally, the spatial temperature variations become more pronounced with increasing instantaneous heat loss. This effect has also been observed in numerical analyses conducted within the scope of this research [Smith 2014].

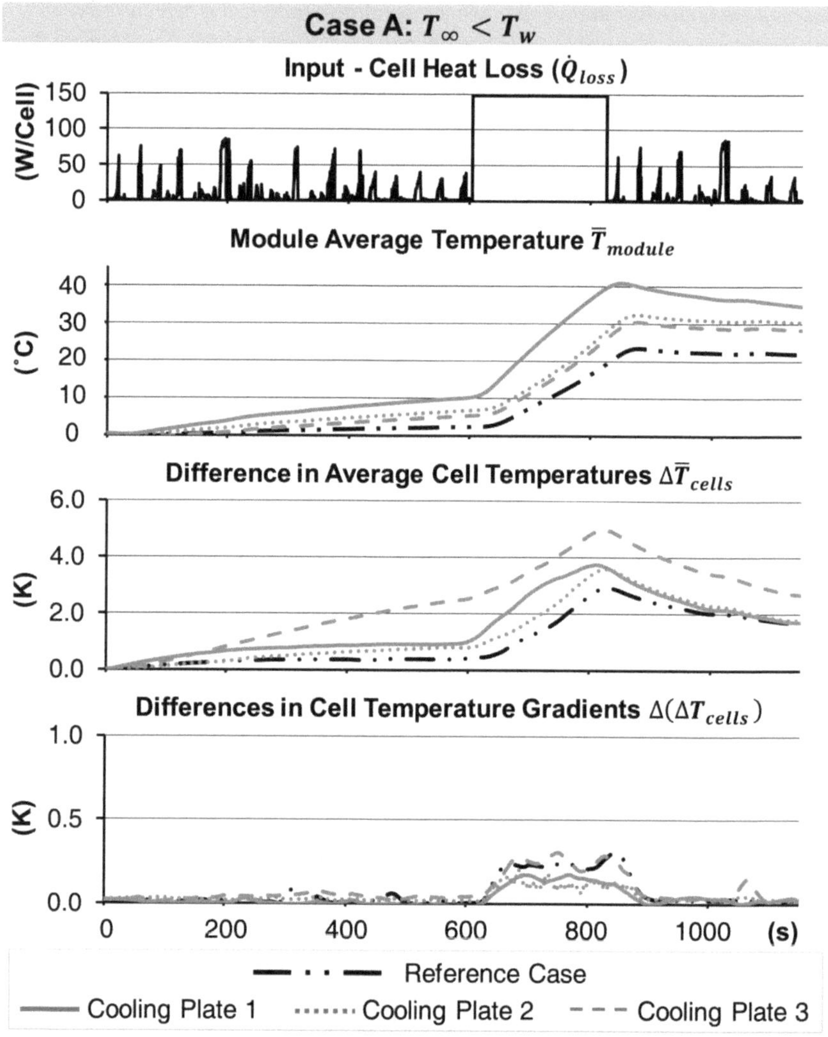

Figure 5.6: The cell het loss profile (input) for case A (T_∞=0°C and T_w=20°C) is shown at top, followed by the three criteria for thermal performance (Section 2.1):, the average module temperature $\bar{T}_{module}$, the difference in average cell temperature $\Delta\bar{T}_{cells}$, and the differences in the magnitude of the cell temperature gradients $\Delta(\Delta T_{cells})$. Data at a constant coolant flow rate (5 L^1min^{-1}) and inlet temperature (20°C) is shown. All plots share the same x-axis.

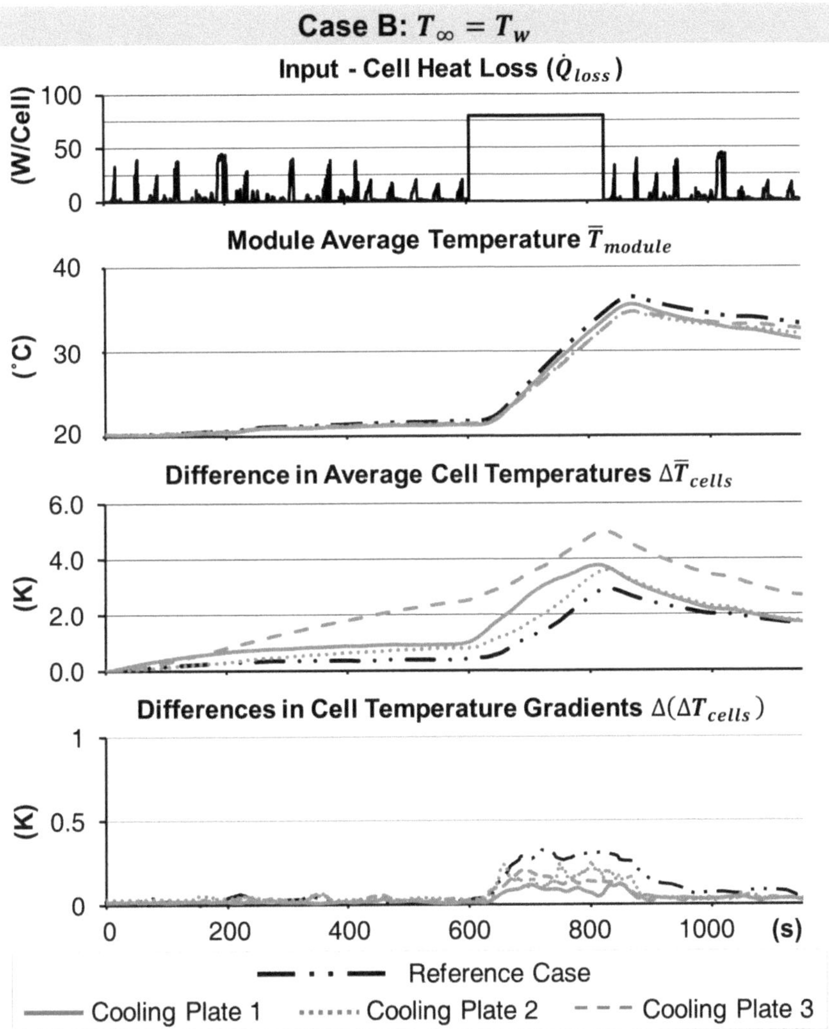

Figure 5.7: The cell het loss profile (input) for case B (T_∞=20°C and T_w=20°C) is shown at top, followed by the three criteria for thermal performance (Section 2.1):, the average module temperature $\bar{T}_{module}$, the difference in average cell temperature $\Delta\bar{T}_{cells}$, and the differences in the magnitude of the cell temperature gradients $\Delta(\Delta T_{cells})$. Data at a constant coolant flow rate (5 L^1min^{-1}) and inlet temperature (20°C) is shown. All plots share the same x-axis.

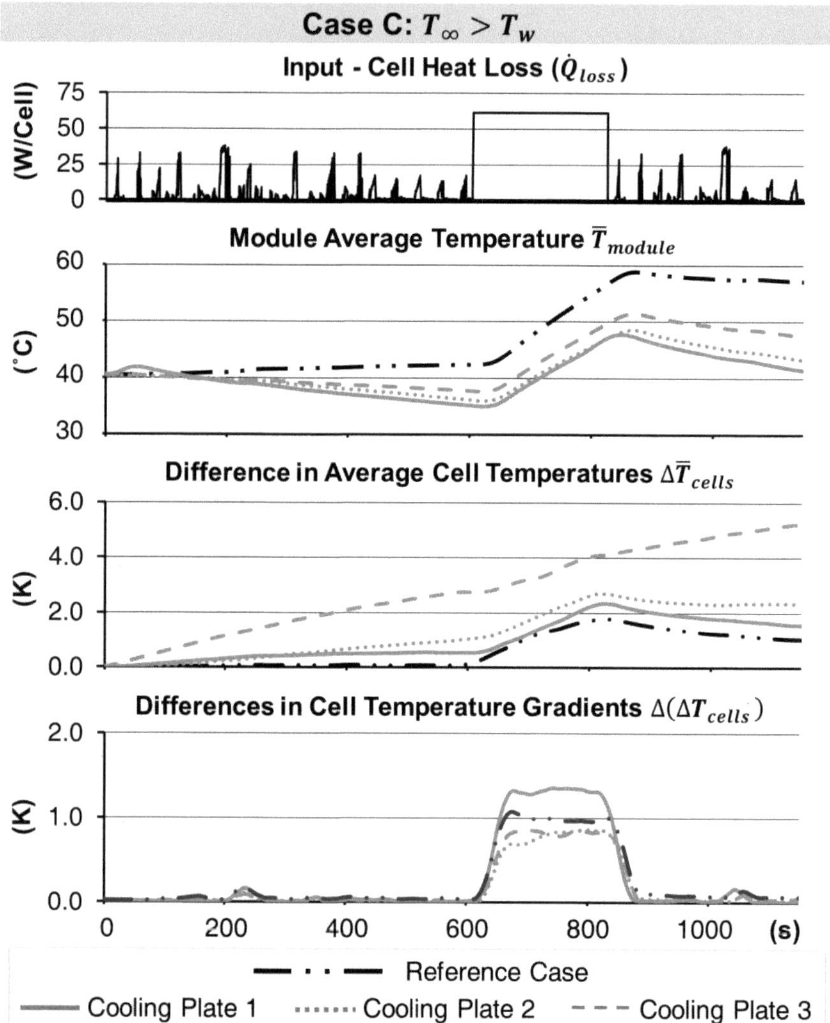

Figure 5.8: The cell heat loss profile (input) for case C (T_∞=40°C and T_w=20°C) is shown at top, followed by the three criteria for thermal performance (Section 2.1), the average module temperature $\bar{\bar{T}}_{module}$, the difference in average cell temperature $\Delta\bar{T}_{cells}$, and the differences in the magnitude of the cell temperature gradients $\Delta(\Delta T_{cells})$. Data at a constant coolant flow rate (5 L^1min^{-1}) and inlet temperature (20°C) is shown. All plots share the same x-axis.

The numerical solution to the first law of thermodynamics is approximated based on the average module temperature. However, due to the addition of the BTMS, Equation 3.3 must be extended to include the heat transfer to/from the BTMS:

$$C\frac{dT}{dt} = \dot{Q}_{loss} - hA_\infty(\bar{T}_{module} - T_\infty) - hA_{BTMS}(\bar{T}_{module} - T_W) \qquad (5.1)$$

where the additional term $hA_{BTMS}(\bar{T}_{module} - T_W)$ includes the thermal conductance (hA_{BTMS}) between the module and coolant with temperature T_W. Because at any time t, the instantaneous cell heat loss is spatially uniform over the module, the mean fluid temperature is used (boundary condition $\dot{Q}_{loss}$ = constant) [Cengal 2007]. The area exposed to the ambient (A_∞) remains the same as for the reference case, and therefore, the thermal conductance to the ambient determined in Section 4.2 is used as the initial guess for fitting the solution. The heat capacity of the cells (C) varies based on cooling plate design (more or less aluminum and/or coolant). The energy input ($\dot{Q}_{loss}$) for cases A through C is identical to the reference case. The cooling plate covers the bottom of the module (A_{BTMS}), which is insulated for the analysis of the reference case. As a result, the additional heat transfer to the cooling plate is expressed by the coefficient hA_{BTMS}, which acts in parallel to the thermal conductance to the ambient. The calculated values are shown in Table 5.2.

Table 5.2: *Iteratively calculated values for the heat capacity (C), the thermal conductance to/from ambient (hA_∞), and the thermal conductance to/from the BTMS (hA_{BTMS}) at the three cases analyzed (described in Section 4). The goodness of fit for of the numerical solution is also included (r^2).*

	Case	C (J^{1}K^{-1})	hA_∞ (W^{1}K^{-1})	hA_{BTMS} (W^{1}K^{-1})	Fit (r^2)
	A	7200	3.0	6.2	0.996
Cooling Plate 1	B	7200	3.2	6.2	0.999
	C	7200	3.0	6.2	0.994
	A	7100	3.0	3.8	0.992
Cooling Plate 2	B	7100	3.2	3.9	0.999
	C	7100	3.2	4.2	0.994
	A	6800	3.0	2.2	0.996
Cooling Plate 3	B	6800	3.2	2.1	0.999
	C	6800	3.3	2.5	0.999

The values of thermal conductance to the ambient (hA_∞) are similar to the fitted solution for the reference case, which agrees with the theory because the boundary condition of natural convection and the exposed surface area are the same. The differences behind the decimal point are assumed to be within the error tolerance of the fitting process. The differences in the values of the heat capacity between designs are a result of the amount of coolant and material in the plate. The initial value for the heat capacity is again rounded, as variations over the considered temperature range are also assumed to be within the error tolerance of the fitting process.

Because all cooling plate technology demonstrators have the same aluminum top plate thickness, the same electrically insulating gap pad, and are compressed with the same force, the coefficient of thermal conduction (h_{BTMS}) between the module and the coolant should be similar. Therefore, the differences in the value of thermal conductance (hA_{BTMS}) between cooling plate designs are a result of different coolant contact areas (A_{BTMS}): the larger the channel footprint A_{BTMS}, the higher the thermal conductivity to the BTMS. This effect can be demonstrated analytically via the Nusselt Number and a thermal resistance network.

The Nusselt Number (Nu) is the dimensionless temperature gradient at the wall, expressing the relationship between the convection into the working fluid (h_w) and the fluid-side conduction (k_w) at the wall of a channel with the hydraulic diameter D. For fully developed laminar flow, the Nusselt Number is constant.

$$Nu = \frac{hD}{k} = constant \tag{5.2}$$

The value of the Nusselt Number depends on the flow channel geometry. From the Nusselt correlation, the coefficient of heat transfer can be calculated based on the properties of the coolant and the hydraulic diameter. The surface temperature of the plate can then be calculated via a thermal resistance network, where the thermal resistance (R) describes the convection from the coolant to the aluminum, and the conduction through the aluminum to the cooling plate surface.

$$T_S = R * \dot{Q} + T_w$$

$$\text{where } R = \left(\frac{1}{hA_{BTMS}} + \frac{t_{Al}}{k_{Al}A_{BTMS}} \right) \tag{5.3}$$

As shown in Equations 5.2 and 5.3, as the BTMS contact surface area (A_{BTMS}) increases, the thermal resistance (R) decreases, and the surface temperature (T_S) approaches that of the coolant. Thus, the effectiveness of the BTMS is increased. The relationship shown is based on laminar flow, which occurs primarily for all cases experimentally analyzed. Transitional or turbulent flow would increase the heat transfer coefficient, but would also increase the pressure drop and thus the energy consumption of the coolant circuit pump (see Equation 5.4).

As presented in Section 2.1, the average temperature of the module is not the only relevant criteria of thermal performance. Large differences in the average temperatures of the individual cells ($\Delta\bar{T}_{cells}$) indicate that the cells are exposed to different temperature levels and thus may age at various rates. The large values of $\Delta\bar{T}_{cells}$ observed especially for cooling plate demonstrator 3 indicate the most significant risk of inhomogeneous aging. Both cooling plate demonstrators 1 and 2 exhibit better performance at high heat loads. As expected, the imposed gradients are larger than for the reference case (two to three degrees); however, the temperature level of the entire module is reduced by over ten degrees, which has a far more beneficial effect on reducing system aging.

The final criteria of inhomogeneous aging, $\Delta(\Delta T_{cells})$, indicates the potential of inhomogeneous aging due to the premature aging of an individual cell caused by an imposed temperature gradient over the cell. All cooling plates imposed similar gradients over the individual cells. The difference is in the absolute value of the temperature difference, which is shown in Table 5.3. Here, cooling plate 1 induces slightly (0.5°C) higher temperature gradients than the other cooling plate designs, which simultaneously have a smaller coolant channel footprint (A_{BTMS}). However, the magnitude of these differences is again within the measurement uncertainty of the thermocouples in the SBC (Section 3), meaning that a larger coolant footprint should be favored.

Hydraulic Work

The hydraulic work, an indication of the energy efficiency of the designs, is compared via the theoretical pump power as presented in Section 2.2. The comparison of theoretical pump power for all inlet and ambient temperatures at flow rates near 5 L^{1}min$^-$^{1} is shown in Table 5.3. Across all conditions tested, cooling plate 2 requires the highest theoretical pump power as a result of the long, serpentine channel, yet the average module temperatures measured for cooling plate demonstrators 2 and 3 are in some cases so similar that they lie within the measurement tolerance of the thermocouples. However, the lower thermal conductance of cooling plate 3 is expressed through the differences in the cell average temperatures at high heat losses. The FDB in cooling plate demonstrator 1 result in a higher pressure drop than the smooth channels of cooling plate demonstrator 3, despite the shorter channel length of the I-flow pattern. In the analysis performed, a case consisting of the best thermal performance and the least hydraulic work do not coincide. Thus, a compromise must be found.

The pressure drop reacts more sensitively to geometrical variations in the cooling plate than the criteria for thermal performance. This experimental observation agrees with the conclusions drawn from previous work using numerical simulations (see Appendix D), and can be explained by the analytical solution for both pressure drop and the heat transfer coefficient for Hagen-Poiseulle flow [Cengal 2007]:

$$\Delta p = 32 \frac{V_{avg}\mu L}{D^2} \tag{5.4}$$

$$h = 4.36 \frac{k}{D} \tag{5.5}$$

where halving the hydraulic diameter quadruples the pressure drop (Equation 5.4) yet only doubles the heat transfer coefficient (Equation 5.5).

The most significant source of the pressure drop is however in most cases not the actual cooling plate, but the connection to the coolant loop. In the cooling plate analyzed, a 90° junction found in many current cooling plate designs was implemented. From pressure loss tables, the equivalent pipe length of a 90° fitting can exceed 60 to 90 times the actual length (L) [Annaratone 2010]. Considering the fluid channel of the cooling plate technology demonstrators, a 90° inlet and outlet each create nearly the same pressure drop as the flow channel within the plate (Equation 5.4). This phenomenon is visualized in Figure 5.9 for a sample cooling plate.

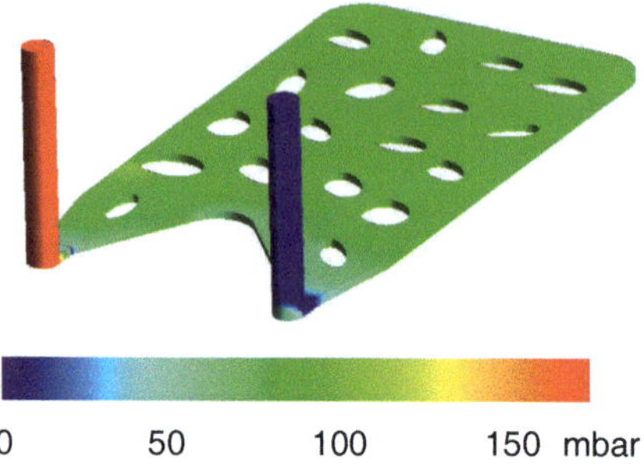

Figure 5.9: Sample contour plot of the pressure drop at the inlet/outlet transitions versus the main flow channel. The plot is intended as a visualization only.

Changing the angle of the inlet/outlet with respect to the plate would alleviate this drop, but possible consequences such as added cooling plate length must be considered in terms of the vehicle suitability.

With respect to the optimum cooling plate design, the small flow channels and consequently the laminar flow mean that the coefficient of heat transfer is only a function of the channel geometry. As such, the channel footprint (A_{BTMS}) must be maximized while ensuring that unnecessary redirection of the flow does not occur. Simultaneously, the hydraulic diameter must be optimizing so that the pressure drop is minimized under consideration of the relationship shown in Equations 5.3 and 5.4.

Vehicle Suitability

The vehicle suitability (Section 2.3) of the cooling plate demonstrators in comparison to one another is shown for both the existing technology demonstrator and for potential future (optimized) layouts in Table 5.3. The "current" values are measured directly, while the "potential" values are based on expert projections for the limits of the joining technology and the minimum flow channel cross-sectional area to reduce the risk of channel clogging. The effect of a minimization of cooling plate size must consider the potentially negative effects on the hydraulic work through a smaller hydraulic diameter and multiple redirections of the flow for example through the inlet and outlets of the cooling plate.

Table 5.3: Multifactorial analysis of the cooling plate technology demonstrators based on the objective evaluation criteria (Section 2.1). The values shown are the average over the cases (A through C) tested.

	Cooling Plate Technology Demonstrator 1		Cooling Plate Technology Demonstrator 2		Cooling Plate Technology Demonstrator 3	
Thermal Behavior						
Thermal Conductance [W^1K^{-1}]	6.2		4.0		2.3	
Heat Capacity [J^1K^{-1}]	7200		7100		6800	
Maximum $\overline{\Delta T}_{cells}$ [K]	3.8		3.6		5.2	
Maximum ΔT_{cell} [K]	1.7		1.9		1.8	
Maximum $\Delta(\Delta T_{cells})$ [K]	1.3		0.9		0.9	
Hydraulic Work						
Theoretical Pump Power[1] [W]	1.2		2.1		0.8	
Vehicle Suitability[2]	*Current*	*Potential*	*Current*	*Potential*	*Current*	*Potential*
Weight: Δm	+ ≈ 12%	+ ≈ 5%[3]	+ ≈ 11%	+ ≈ 5%[3]	+ ≈ 4%	+ ≈ 3%[4]
Volume: ΔV	+ ≈ 9%	+ ≈ 9%[3]	+ ≈ 8%	+ ≈ 8%[3]	+ ≈ 5%	+ ≈ 3%[4]
Length: ΔL_x	+ ≈ 18%	+ ≈ 10%	+ ≈ 9%	+ ≈ 9%	+ ≈ 16%	+ ≈ 9%
Width: ΔL_y	+ ≈ 0%	+ ≈ 0%	+ ≈ 0%	+ ≈ 0%	+ ≈ 0%	+ ≈ 0%
Heigth: ΔL_z	+ ≈ 5%	+ ≈ 5%	+ ≈ 5%	+ ≈ 5%	+ ≈ 8%	+ ≈ 8%

[1] at a flow rate of 5 L^1min^{-1}, averaged over all trials

[2] in comparison to an 8-cell module with no thermal management, excluding the mechanical bracing

[3] depending on the necessary structural layout and material combinations, full coverage of module base

[4] partial coverage of module base and no structural function

Producibility and Economic Viability

For the implementation of larger coolant channels, thicker channel walls are required to minimize deformation due to the internal pressure. Correspondingly, joining techniques providing additional strength are required as the absolute pressure within the channel increases. Variations of FSW are thus advantageous, as material pairings of different thickness, alloy, or type can be combined to achieve the necessary structural support and simultaneous structural integration within the vehicle.

Soldering provides the opportunity for the lightest-weight part because thin material profiles can be joined easily. For applications were a functional integration of cooling plate structure is not foreseen, soldered aluminum profiles are currently the best mass production solution due to experience of manufacturers with vehicle coolers.

Excellent results were also obtained via adhesion. The main drawback are the surface preparation and the curing time; however, if the curing time can be compensated or is not relevant due to the production strategy, adhesion can be applied cost effectively and universally.

As shown in Section 2.4, to reduce the total lifecycle impact of a cooling plate, the amount of raw material should be minimized by, for example, designing the part to use recycled aluminum, or through the use processes better suited to thin-walled material. However, due to the relatively small mass of a thermal management system relative to the total vehicle, the absolute reduction in life cycle energy consumption may be minimal if not negligible.

Summary of the Multifactorial Analysis

The analysis of cooling plate shows the broad range of layouts possible. The optimal cooling plate design, however, is a function of the specific requirement of each vehicle type. The primary conclusions from the analysis are:

- Coolant channel footprint (A_{BTMS}) has the greatest influence on the heat transfer to/from the module;
- The temperature distribution across the module ($\Delta \bar{T}_{cells}$) cannot be neglected when maximizing heat transfer and minimizing pressure drop;
- Flow displacing bodies can be used to locally alter / enhance heat transfer;
- An I-Flow layout causes larger gradients over individual cells near the fluid inlet as opposed to those near the outlet;
- A U-Flow can minimize the temperature differences over the module, but the channel footprint (A_{BTMS}) must be large enough to accommodate the heat transfer for the expected use case;
- The connection of the inlet and outlet to the cooling plate present a significant source of pressure drop. Optimizing this junction cannot be neglected;
- The amount of solid material should be minimized to meet the constraints of the chosen production techniques and the structural requirements of the cooling plate, both from external mechanical loads and the internal pressure resulting from the coolant.

Based on the conclusions from the multifactorial analysis, recommendations are made for cooling plate design for battery thermal management in a high-performance PHEV.

5.3. Discussion and Design Recommendations

Based on the evaluation criteria of this research, increasing cooling plate channel footprint (A_{BTMS}) serves to reduce pressure drops and facilitate maximum heat transfer. The integration of fluid displacing bodies can be used to locally enhance heat transfer and improve temperature homogeneity, while providing structural support and a joining surface. The U-flow layout allows for the inlet and outlet to be positioned on the same side, resulting in additional volume on only one side of the cooling plate, which better suits vehicle integration in many cases. Ideally, the inlet and outlet angle in respect to the cooling plate would be reduced, as the junction caused by a 90° angle between inlet/outlet and cooling plate creates significant pressure drops with respect to the cooling plate itself. This conclusion applies to all coolant connections and circuitry in the vehicle: additional length and bends increase the pressure drop. Additionally, cooling plate design must occur in conjunction with the layout of the thermal interface between cells and cooling plate. This layer must minimize thermal resistance in order to increase the efficiency of the cooling plate. Based on the multifactorial analysis, two design suggestions are made for the cooling plate of a luxury, high performance PHEV, depending on the degree of structural integration.

No Structural Integration

For a lightweight cooling plate that is not required to bear any load beyond the cell module, the state-of-the-art design using soldered aluminum profiles is unmatched. The layout can be optimized to meet the necessary thermal performance, and the production process is well proven from the vehicle cooler industry. The thin aluminum profiles can withstand some compression force, allowing for increasing forces between module and BTMS. The profiles can even be compressed plastically to ensure better contact; however, a vehicle service concept requiring module replacement is then difficult, and channel deformation resulting in undesired pressure drop and altered thermal performance may result.

Additionally, the height of a cooling plate produced in this way is channel clogging: small channels ($\approx$0.5 mm) tested required careful filtering of the coolant. Therefore, a minimum channel height of 1 mm is recommended to guarantee robust performance. Because of the availability of extruded aluminum pieces in many sizes, this method would be suitable to lower production volumes as well, especially when combined with an existing vehicle heat exchanger production.

Structural Integration

For structural integration within the vehicle chassis, a cast or deep-drawn battery casing designed to contribute to the vehicle structure and house the battery modules is recommended. The battery modules can be compressed to the casing floor (with an electrically insulating layer between) to enhance the thermal contact interface. A polymer or metal plate can be joined to the casing from the outside via FSW, EBW, or adhesion, as shown in Figure 5.10. In this way, any failure in the joint between components results in coolant leakage outside the battery casing.

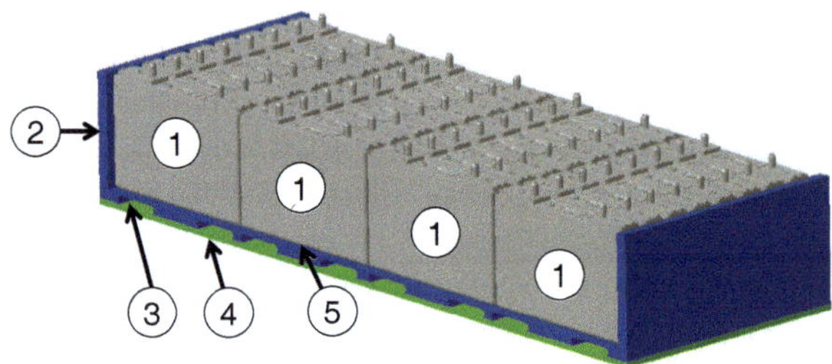

Figure 5.10: Cross-section of the proposed layout for a structurally integrated cooling plate. Shown are module (1), a cast or deep-drawn casing (2) with integrated channels (3), and a polymer or metal bottom plate (4) joined via FSW or adhesion. The modules are pressed over a thermal interface (5) onto the casing floor.

In this design, the casing should provide a two-dimensional joining surface (versus three-dimensional) so that a *planar* piece of material can be attached. Achieving leak-tightness between two components having surfaces varying in three-dimensions (versus flat, two-dimensional surfaces) proved challenging in the scope of this work. Casting and deep-drawing do however require higher production volumes to become economically viable, but given the nature of the upcoming CO_2 regulations, a structurally integrated battery casing as part of a vehicle platform for multiple models would presumably meet this criterion. The functional integration of a battery casing within the vehicle structure provides the potential for total vehicle weight savings.

6. Investigation of Active Spacers

In this section, an alternative thermal contact surface is considered: the space between the cells, within a module. In this case, so-called active spacers having extruded channels streamed with water-glycerol are investigated. The thermal contact area between a spacer and an individual cell is much higher than between a cooling plate and cell: with an active spacer between each PHEV-2 cell, the contact area increases seven-fold. Additionally, more of the cell jelly-roll (Figure 3.4) surface is contacted, which results in lower thermal resistance between the BTMS and the active chemistry of the cell. In terms of vehicle suitability, the spacers add length to the module but reduce height in comparison to a cooling plate. Active spacers are widely applied for pouch cells, but no production application implementing them for PHEV-2 cells is known to the author.

Two types of primary materials are considered for the actively-streamed spacer: ceramic and aluminum. Because aluminum was investigated experimentally via the cooling plate technology demonstrators (Section 5), ceramic spacers are utilized in the experimental feasibility study, while aluminum is compared numerically. Ceramic is an intriguing material not only for use with high-voltage batteries, but in the power-electronics industry in general. Ceramics, depending on their exact make-up, have a broad spectrum of properties. They can be electrically insulating or conducting, thermally insulating or conducting, and have varying degrees of harness and strength [Huelsenberg 2014]. Ceramics are ideal for applications involving high mechanical and thermal stresses where corrosion resistance is required, such as a spacer between battery cells [Michaelis 2009].

For the spacers between battery cells analyzed in this Section, an electrically insulating ceramic is considered in order to eliminate additional electrically insulating layers, as are required for aluminum spacers. Additionally, a thermally conductive ceramic is chosen to further reduce the thermal resistance from the cells to the BTMS. Given the high compressive forces generated between the considered PHEV-2 cells, the mechanical properties of ceramic are ideal to counter this force and reduce cell deflection. Various surface finishes are available to better enhance thermal contact. The primary disadvantages of ceramic versus aluminum is the high density ($\approx$30%) and therefore generally weight, as well as the higher material cost.

Ceramic production and the layout of the technology demonstrator is presented in the following section. In Section 6.2., the experimental data and multifactorial analysis are shown. Section 6.3. provides a summary of the complete analysis including design recommendations and suggestions for improvement.

6.1. Layout and Considered Production Techniques

Ceramics differ from metals in that they generally have a negative temperature coefficient, meaning that electrical resistivity lowers with increasing temperature [Huelsenberg 2014, Hornbogen 2008]. In comparison to polymers, ceramics do not depend on a weakly-linked network of chains, but rather on an ordered structure of stable covalent and ionic bonds that given ceramics the hardness, high modulus of elasticity, and electrical insulation they are most known for [Hornbogen 2008]. Nearly all ceramics are manufactured by power forming, which involves densifying the powder via solid state sintering at approximately 70 to 90% of the material's melting point [Ruys 2015]. Ceramic is however formed in its so-called "green" state (pre-sintering). The forming process, which can involve wet pressing, injection molding or slip-casting to name a few, must consider the dimensional changes (up to 10%) that occur during sintering [Ruys 2015]. Due to the crystalline structure of ceramics, the packing density is difficult to control during pressing, especially compared to metals; as a result, thinner extruded profiles and a greater control of mechanical properties can be realized with aluminum or other metals [Ruys 2015, Ghassemali 2015].

Many technical ceramics meeting the aforementioned criteria (electrical insulation and thermal conduction) contain at least some Aluminum-Oxide [Huelsenberg 2014]. The high compressive strength (2000 to 4000 MPa), the good thermal conductivity (20 to 30 $W^1m^{-1}K^{-1}$), high electrical insulation ($1x10^{14}$ to $1x10^{15}$ Ωcm) and exceptional corrosion resistance speak for the material [Ceramtec 2015]. Another interesting material that has grown in popularity since its market introduction 25 years ago is a non-oxide ceramic, Aluminum-Nitride (AlN). With thermal conductivity a full order higher than Aluminum-Oxide ($\approx$200 $W^1m^{-1}K^{-1}$) and sufficiently high electrical insulation ($1x10^{12}$ Ωcm), this material shows promise for thermal management applications [Huelsenberg 2014, Ceramtec 2015]. However, the production is more challenging, requiring protective gasses and higher sintering temperatures [Huelsenberg 2014].

In either case, the very low coefficient of thermal expansion allows metal plating or screen printing of thin (2 µm) Nickel on the material surface [Huelsenberg 2014, Michaelis 2009]. As a result, ceramics have found wide acceptance in the semiconductor and power electronics industries. In the case of a BTMS, plating can be used to provide a surface for joining components to the ceramic via soldering or to apply a heating band directly on the spacer.

For the technology demonstrator, extruded ceramic profiles approximately 3 mm thick with 1.8 mm wide flow-channels are metalized at the ends to facilitate soldering with copper end-pieces, which are then soldered to a T-junction.

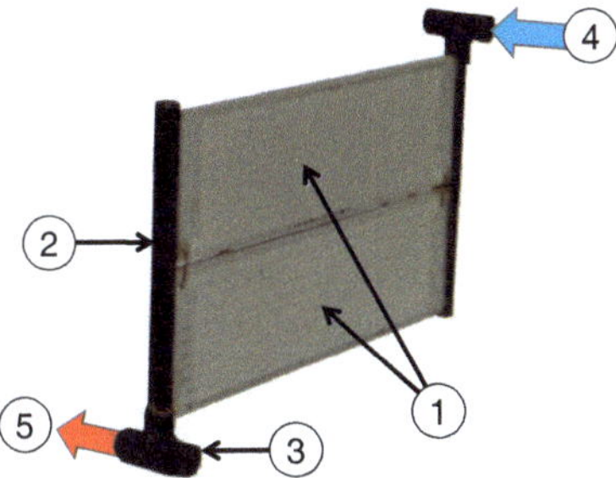

Figure 6.1: A ceramic spacer utilized for the feasibility study consisting of two extruded ceramic profiles (1), a copper distributer-tube on each side (2), and a T-junction to connect spacers (3). The inlet (4) and outlet (5) for each spacer are shown.

The ceramic thickness as well as the multi-piece layout were chosen because this standard geometry was available from the supplier: specifically developing a ceramic profile would have inflated feasibility study costs. As a feasibility study, the goal is to first access the functionality of the concept. Spacers of smaller thicknesses and alternative channel profiles were analyzed via simulation as an input for future design iterations. The copper tubes and T-junctions were also chosen because they could be easily welded to the ceramic via the nickel plating by hand: copper would be ill-suited in a production version because of the cost and weight versus aluminum or a polymer joined via another technique.

The spacers are connected to each other via small copper tube sections, allowing individual spacers to be removed so that alternative layouts can be investigated. This simple method proved leak-tight for the experiments, but such a rigid connection is not recommended for a mass-production version because of the low tolerance flexibility.

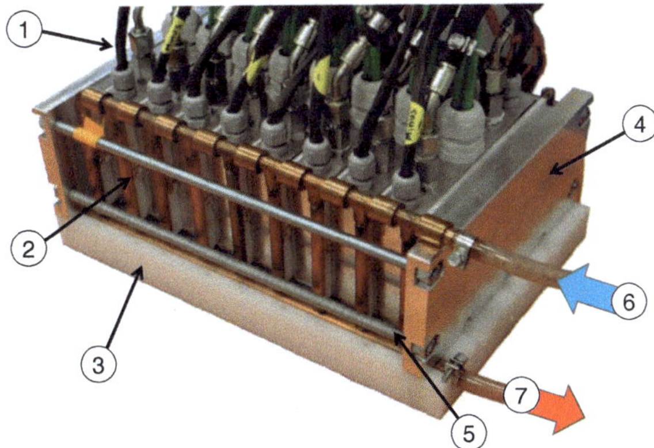

Figure 6.2: The complete technology demonstrator including spacers (1) connected by copper tubing sections (2), the thermally insulating base (3), the compressive endplates (4), the carriage bolts (5) as well as the system inlet (6) and outlet (7).

The position of the inlet and outlet were chosen to be on one side in order to facilitate a better integration with the testing equipment and present a connection situation similar to the cooling plates (U-flow). The inlet and outlet position and geometry of each spacer is the same to simplify production. The demonstrator was experimentally investigated in three configurations: a spacer between each cell, a spacer between each second cell, and a spacer between each fourth cell (every third cell was not investigated because an eight-cell module was used). The goal is to minimize the amount of spacers required to achieve an acceptable thermal performance, as additional spacers add size, weight and cost to the system. When an active spacer is removed from the technology demonstrator, a polycarbonate spacer—just as used in the cooling plate technology demonstrators—replaces it. The experimental data and multifactorial analysis are presented in the following section.

6.2. Multifactorial Analysis of Ceramic Spacers

An active spacer between each PHEV-2 cell increases the thermal contact area seven-fold over the use of a cooling plate, every second cell still results in three-and-a-half times the contact area. Additionally, the thermal resistance to the cell jelly-roll and therefore the active chemistry is much smaller, so that active spacers can potentially have a greater effect on reducing premature cell aging or temperature gradients over and between cells. However, the layout involving a spacer every fourth cell leaves multiple cells without direct thermal contact to the BTMS, and significantly effects the thermal performance of the module.

The thermal performance and theoretical pump power are again experimentally measured at various ambient temperatures, coolant flow rates, and coolant inlet temperatures using a 50/50 water-glycerol. In all cases, the boundary conditions are held constant throughout the trial. To facilitate clarity in the plots throughout the research, the case of 20°C coolant inlet temperature and 5 L^1min^{-1} flow rate is shown. As expected, a lower coolant inlet temperature and higher flow rate decreases module average temperature. This trend is consistent over all trials. The multifactorial analysis is presented below.

Thermal Performance

The thermal performance of the technology demonstrators is evaluated based on the criteria presented in Section 2.1, and plotted for the three cases considered in Figures 6.3 through 6.5.

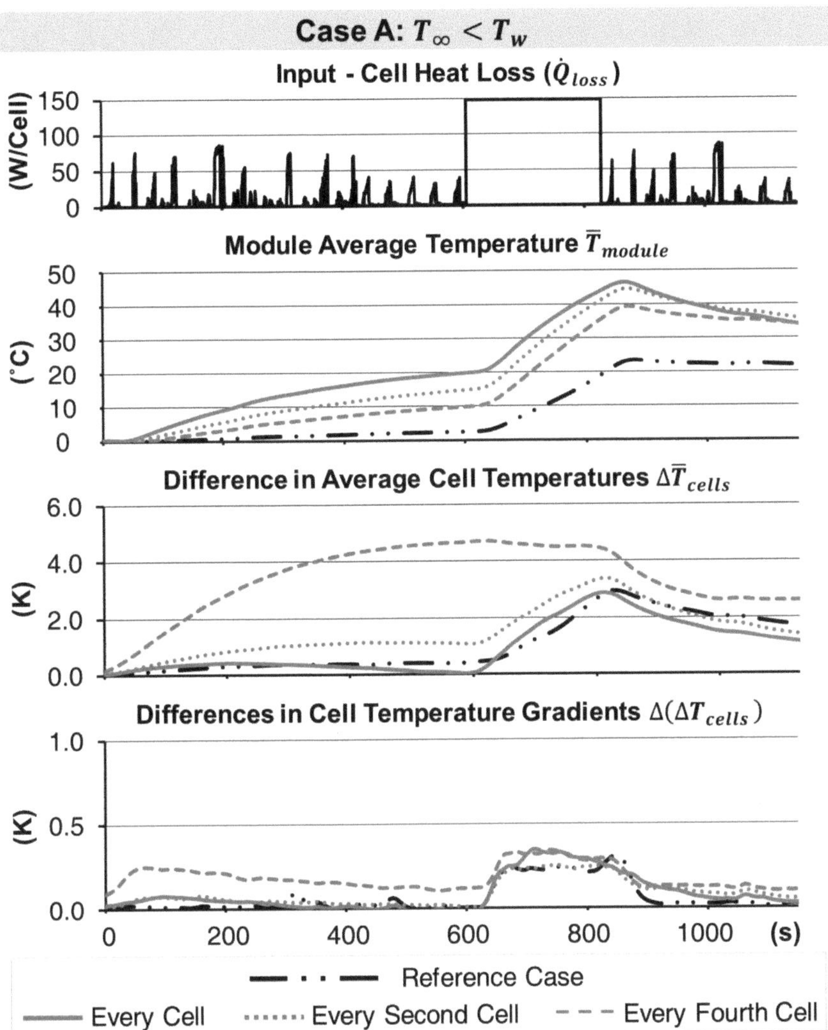

Figure 6.3: The cell het loss profile (input) for case A (T_∞=0°C and T_w=20°C) is shown at top, followed by the three criteria for thermal performance (Section 2.1):, the average module temperature $\bar{T}_{module}$, the difference in average cell temperature $\Delta\bar{T}_{cells}$, and the differences in the magnitude of the cell temperature gradients $\Delta(\Delta T_{cells})$. Data at a constant coolant flow rate (5 L^1min^{-1}) and inlet temperature (20°C) is shown. All plots share the same x-axis.

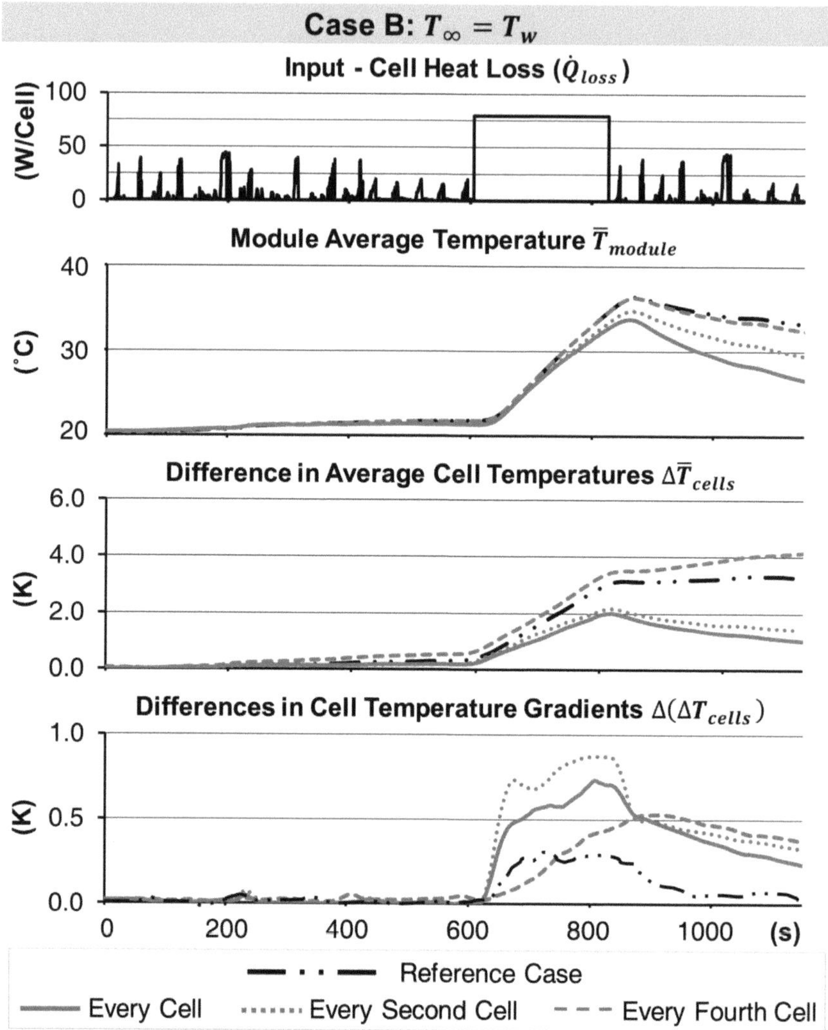

Figure 6.4: The cell het loss profile (input) for case B (T_∞=20°C and T_w=20°C) is shown at top, followed by the three criteria for thermal performance (Section 2.1):, the average module temperature $\bar{T}_{module}$, the difference in average cell temperature $\Delta\bar{T}_{cells}$, and the differences in the magnitude of the cell temperature gradients $\Delta(\Delta T_{cells})$. Data at a constant coolant flow rate (5 L^1min^{-1}) and inlet temperature (20°C) is shown. All plots share the same x-axis.

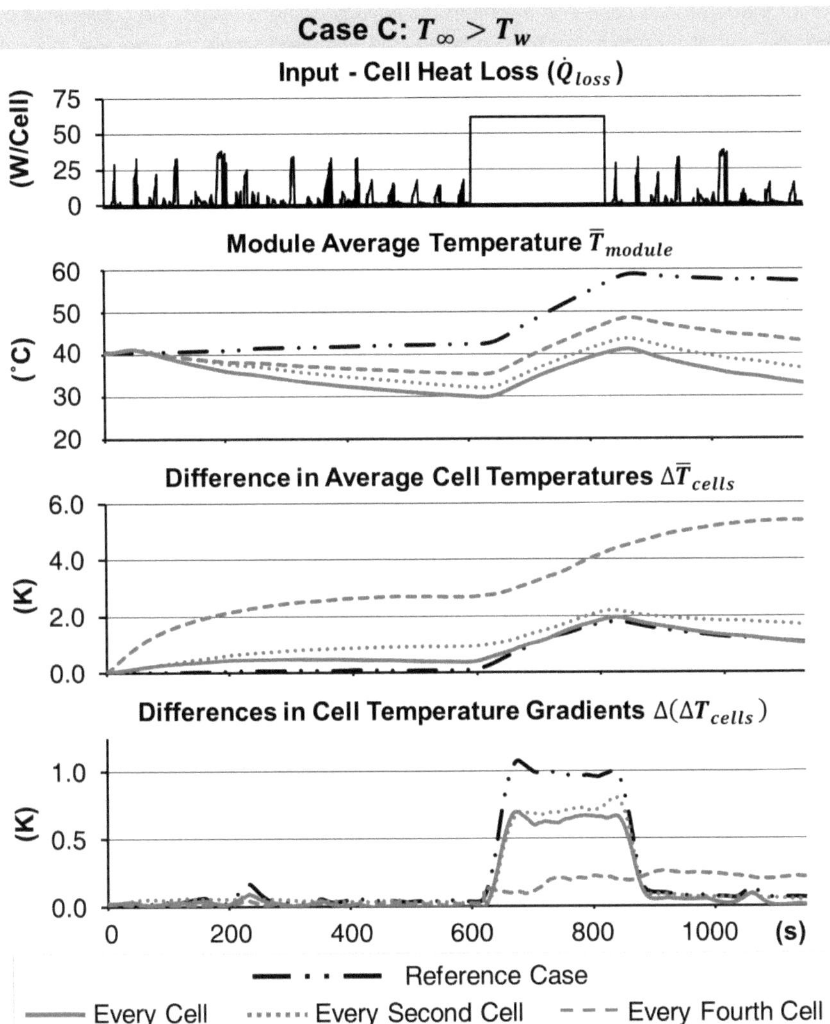

Figure 6.5: The cell het loss profile (input) for case C (T_∞=40°C and T_w=20°C) is shown at top, followed by the three criteria for thermal performance (Section 2.1):, the average module temperature $\bar{T}_{module}$, the difference in average cell temperature $\Delta\bar{T}_{cells}$, and the differences in the magnitude of the cell temperature gradients $\Delta(\Delta T_{cells})$. Data at a constant coolant flow rate (5 L^1min^{-1}) and inlet temperature (20°C) is shown. All plots share the same x-axis.

As for the reference case and the other technology demonstrators, the coefficients of the first law of thermodynamics (Equation 5.2) are fitted using the average module temperature. The area to the ambient (A_∞) is the same as in the reference case and the for the cooling plate demonstrators; therefore, the thermal conductance from the reference case is used as an initial guess for the iterative fitting of the coefficients. The fitted thermal conductance values as well as the heat capacities are shown in Table 6.1. The calculated heat capacities are again used as an initial guess. The fitted heat capacities are rounded to the nearest hundred because further accuracy is assumed unrealistic for the fitted solution. Additionally, the variations in thermal conductance observed (one to two tenths) are considered to fall within the error tolerance of the fitting method.

Table 6.1: Iteratively calculated values for the heat capacity (C), the thermal conductance to/from ambient (hA_∞), and the thermal conductance to/from the BTMS (hA_{BTMS}) at the three cases analyzed (described in Section 4). The goodness of fit for of the numerical solution is also included (r^2).

	Case	C (J^{1}K^{-1})	hA_∞ (W^{1}K^{-1})	hA_{BTMS} (W^{1}K^{-1})	Fit (r^2)
Every Cell	A	7500	3.2	19.4	0.996
	B	7500	3.2	19.5	0.998
	C	7500	3.2	19.5	0.997
Every Second Cell	A	7100	3.2	10.9	0.998
	B	7100	3.2	10.9	0.998
	C	7100	3.1	10.8	0.995
Every Fourth Cell	A	6900	3.0	5.5	0.998
	B	6900	3.2	5.6	0.998
	C	6900	3.2	5.7	0.997

As expected, the thermal conductance is greatest when more spacers are used, as the contact area to the BTMS (A_{BTMS}) is larger. This effect is shown through the nearly linear decrease in thermal conductance as the number of spacer are reduced between layouts from nine to five, then from five to three. The rate at which the module average temperature approaches the coolant temperature is also highest when more active spacers are used: this is evident when viewing the slopes of the average temperature curve in Figure 6.3, and when considering the relationship between heat capacity and thermal conductance (the time constant). The value of the fitted coefficients of thermal conductance to both the ambient and coolant are most sensitive to changes at low cell heat loss ($\dot{Q}_{loss}$).

Not only the average module temperatures between the various spacer layouts differs, but also the spatial temperature variations increase drastically when every spacer is not in contact with the BTMS, as shown for the layout of a spacer every fourth cell in Figures 6.3 through 6.5. However, for the layout of a spacer between each cell and each second cell, for both of which each cell is in contact with the BTMS, the differences in the cell average temperature ($\Delta \bar{T}_{cells}$) are insignificant.

The significant temperature differences amongst cell due to the layout of a spacer every fourth cell will drastically effect system lifetime. The origin of the resulting temperature gradients is analyzed in more detail in Figure 6.6. The results show that not only the lack of BTMS contact surface to each cell, but also the module-level flow pattern effect the temperature gradients.

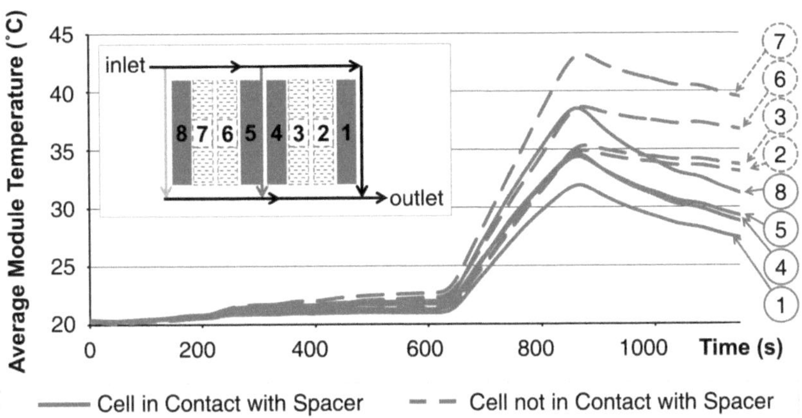

Figure 6.6: Analysis of a module with one spacer every fourth cell shown for case B (20°C ambient temperature, 20°C coolant inlet temperature and 5 L¹min⁻¹ flow rate). The cells are numbered in the plot and the schematic of the module (view from above, top left of plot). In the schematic of the module, arrows show the direction of coolant flow and relative flow rate, with a lighter arrow indicating a lower flow rate.

As shown in Figure 6.6, the cells with the most optimal average temperature have contact with an active spacer. However, the differences between the average temperature of these cells (1, 4, 5, 8) is also fairly significant (4 to 5°C), indicating that the flow coolant flow to the spacers is not homogeneous. In future optimizations of such a technology demonstrator, system-level flow must be considered, as shown by this data. These results are also visualized via numerical simulations in Appendix E.

The maximum temperature difference over a single cell is high for each layout (Table 6.2), because the high heat transfer to the spacer which indicates the potential for aging; however, in all layouts, the differences between the gradients ($\Delta(\Delta T_{cells})$) are similar and within the measurement uncertainty of the thermocouples within the SBC (Section 3), indicating *homogenous* aging over all the cells.

An optimized flow pattern is essential to ensuring the uniform distribution of the coolant and thus a homogeneous temperature throughout the module. Even in the form of a technology demonstrator, the layouts of a spacer every cell and a spacer every second cell improve or match the difference between the average temperatures of the individual cells ($\Delta \bar{T}_{cells}$) versus the reference case, while significantly lowering the module average temperature.

Hydraulic Work

Pressure drop over the total BTMS increases as the number of spacers in the module decreases, because the same flow rate is distributed over a decreasing number of channels and a higher local flow velocity results (Table 6.2). In comparison to the smooth channels within the spacers, the connection pieces and inlet/outlets between the spacers dominate the pressure drop [Annaratone 2010].

For vehicle integration, minimizing spacer thickness is desirable. As a result, the channel thickness is also reduced. Additionally, reducing the flow rate reduces the pressure drop, as shown in Table 6.2 under hydraulic work. This relationship is characterized by Equations 5.4 and 5.5: reducing the hydraulic diameter of the channel by half increases pressure drop four-fold. This sensitivity to pressure drop is contrasted with the effect of a smaller channel on the average module temperature: the variation in the temperature at the cell surface (T_s) is minimal for the spacer thickness considered. As for the case of the cooling plate (Equation 5.3), a thinner spacer (and thus thinner wall) lowers the thermal resistance between coolant and spacer surface. As a result, the spacer surface temperature approaches that of the coolant. The sensitivity of pressure drop to channel thickness versus surface temperature is shown in Figure 6.7 for three spacer thickness.

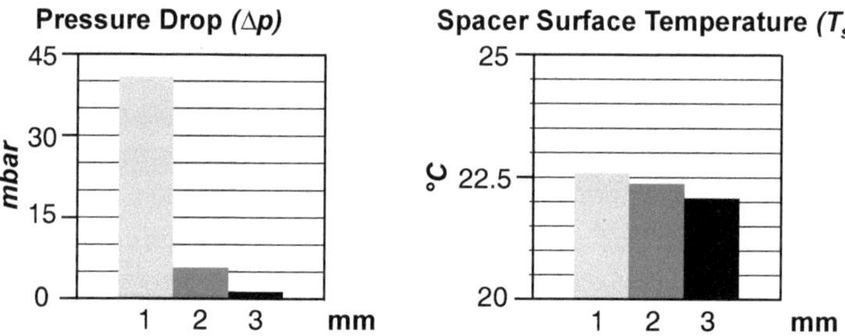

Figure 6.7: The pressure drop (Δp) and the surface temperature (T_s) on a 1 mm, 2 mm and 3mm spacer between each cell. Shown is the steady state data for an ambient temperature of 20°C, coolant inlet temperature of 20°C, and a flow rate of 2 L¹min⁻¹. The aspect ratio of the channels is maintained while scaling.

Figure 6.7 shows the large increases in pressure drop caused by a decreased channel cross sectional area and the resulting increase in flow speed. Correspondingly, the coefficient of heat transfer increases (Equation 5.5) so that the spacer surface temperature (T_s) approaches that of the 20°C coolant. The differences in surface temperatures between spacer thicknesses are far smaller than the differences in pressure drop, highlighting the importance of considering the pressure drop, as the effect on thermal management system energy consumption can be significant. These findings correspond to numerical simulations performed within the scope of this research (Appendix D).

Vehicle Suitability

The vehicle suitability (Section 2.3) of the ceramic spacer layouts in comparison to one another is shown for both the existing technology demonstrator and for potential future (optimized) layouts in Table 6.2. The current values are directly measured, while the potential layouts are in this case based on an aluminum extruded profile, as thinner spacers are possible (0.5 versus 1 mm). The vehicle suitability is enhanced by using a smaller spacer as well: by using a 1 mm spacer instead of a 3 mm spacer, an eight-cell module is smaller (18 mm or 7.5% shorter) and lighter (846 g or 11%).

Table 6.2: Multifactorial analysis of the ceramic spacer technology demonstrators based on the objective evaluation criteria (Section 2.1). The values shown are the average over the cases (A through C) tested.

	Spacer between each Cell		Spacer each Second Cell		Spacer each Fourth Cell	
Thermal Behavior						
Thermal Conductance [W^1K^{-1}]	19.5		10.9		5.6	
Heat Capacity [J^1K^{-1}]	7500		7100		6900	
Maximum $\overline{\Delta T}_{cells}$ [K]	2.9		3.4		5.4	
Maximum ΔT_{cell} [K]	2.0		1.8		1.9	
Maximum $\Delta(\Delta T_{cells})$ [K]	0.7		0.8		0.5	
Hydraulic Work						
Theoretical Pump Power[1] [W]	5.4 $(0.41)^2$		5.9 $(0.5)^2$		6.0 $(0.53)^2$	
Vehicle Suitability[3]	*Current*	*Potential*	*Current*	*Potential*	*Current*	*Potential*
Weight: Δm	+ ≈ 21%	+ ≈ 7%[4]	+ ≈ 12%	+ ≈ 4%[4]	+ ≈ 7%	+ ≈ 1%[4]
Volume: ΔV	+ ≈ 14%	+ ≈ 6%[4]	+ ≈ 8%	+ ≈ 3%[4]	+ ≈ 5%	+ ≈ 2%[4]
Length: ΔL_x	+ ≈ 14%	+ ≈ 4%	+ ≈ 8%	+ ≈ 2%	+ ≈ 5%	+ ≈ 1%
Width: ΔL_y	+ ≈ 7%[5]	+ ≈ 3%[5]	+ ≈ 7%[5]	+ ≈ 3%[5]	+ ≈ 7%[5]	+ ≈ 3%[5]
Heigth: ΔL_z	+ ≈ 0%	+ ≈ 0%	+ ≈ 0%	+ ≈ 0%	+ ≈ 0%	+ ≈ 0%

[1] *at a flow rate of 5 L^1min^{-1}, averaged over all trials*
[2] *at a flow rate of 2 L^1min^{-1}, averaged over all trials*
[3] *in comparison to an 8-cell module with no thermal management, excluding the mechanical bracing*
[4] *using aluminum extruded spacers with soldered or glued distributer at end*
[5] *including the optimization of the distributer channel routing*

Producibility and Economic Viability

In regards to production complexity and current material costs, an aluminum spacer is beneficial. Aluminum spacers can be directly soldered (as in cooling plate technology demonstrator 3), providing a stable production process and leak-tightness for water-glycerol or refrigerant. The electrical insulation would in this case be solved by a thin (e.g. 0.025 mm Kapton®) film on the spacer surface as used in on circuit boards. The relatively smooth surfaces of the spacer and cell casing, combined with the large compression forces between cells facilitate good thermal contact. The spacer must, however, be able to withstand these forces without leakage or failure. In this respect, ceramic is advantageous due its strength under compressive loading: the channels could be larger, thereby increasing the channel footprint. Because the ceramic spacer must also be plated to attach a metal distributer piece or glued to use a polymer distributer, the number of production steps required are similar to the application of an electrically insulating film to an aluminum spacer that was soldered to a distributer piece in one step. In consideration of the life cycle energy consumption, ceramic is advantageous versus primary aluminum, as shown in Section 2.4.

Summary of the Multifactorial Analysis

Considering the thermal behavior and the theoretical pumping power of the various systems, utilizing a spacer between every second cell provides adequate thermal performance while saving material and weight versus a module with a spacer between each cell. Furthermore, fewer connection points result in fewer assembly steps and therefore fewer sources of potential leakage. The temperature gradients resulting from a spacer every fourth cell are too detrimental to system lifetime, especially with the reference spacers (3 mm polycarbonate) replacing the active spacers, as in the experimental technology demonstrator. The primary conclusions from the analysis are:

- Thermal contact to the BTMS must be realized for all cells (or none): the temperature gradients created by contact with only certain cells increase the risk of inhomogeneous aging versus the reference case;
- Flow channel cross sectional area affects the thermal performance more significantly than the solid material for the spacer sizes analyzed;
- Ceramic is interesting as a material due to the structural resistance of the compressive forces of the cells and guaranteed electrical insulation, yet the weight and current cost limit its application as a spacer material.

Based on the conclusions from the multifactorial analysis, recommendations are made for the use of water-glycerol streamed spacers and the implementation of ceramic in general for battery thermal management in a high-performance PHEV.

6.3. Discussion and Design Recommendations

This technology demonstrator and the derivations thereof were designed firstly to explore the effect of thermal management system contact within a module (i.e. between prismatic cells), and secondly to evaluate the implementation of ceramic in a BTMS.

Thin (< 1mm) aluminum spacers are best suited for an ultra-high performance/race track application, as they present the best thermal performance at the lowest weight (30% less than ceramic). For performance use, a spacer every second cell is the ideal configuration in terms of thermal performance, energy consumption, vehicle suitability, production complexity and economic viability. In this way, each cell is still in direct contact with the BTMS (versus a spacer every fourth cell). The use of spacers is also attractive for vehicle concepts limited in height, as the cell is the highest part. Thin extruded aluminum profiles may deform plastically under the high compressive loads. This may facilitate a better thermal contact, but channel blockage or even leakage may be possible as a result. Soldered multiple-piece designs having a solid aluminum plate on either side of the flow channel may prevent deformation, but a thickness below 3 to 4 mm is then not feasible. The light-weight and relatively low material cost speak for aluminum; however, currently available electrically insulating films developed for the consumer electronics market are expensive, especially considering the quantities required for a battery system.

The primary benefit of ceramic is the ability to withstand large compressive forces without deformation: as a result the cell shape can be maintained, and the spacers will not deform in vehicle operation, allowing for a bad cell to be replaced in a vehicle. The material is however relatively heavy, and a 1 mm thick spacer is at the limit of current conventional production technologies. With the growing prevalence of additive production techniques using ceramic, thinner profiles may be possible. The most promising use for ceramics given the current price and weight seems to be for a cooling plate application. Using the 3 mm thick profiles, a simple U-flow cooling plate (e.g. cooling plate technology demonstrator 3, Section 5) would provide similar thermal performance to the aluminum version without the need for electrically insulating layers. A high compressive force between the cells and module could also be realized.

7. Investigation of Heat Pipe Spacers and Long Heat Pipes

A heat pipe can transfer heat at orders of magnitude higher than a solid metal without the use of moving parts [Dunn 1982, Noie 2014, Singh 2007]. Very high heat loads can be transferred while maintaining a small temperature gradient due to the high effective conductivity [Dunn 1982]. In a properly laid out heat pipe, functionality can be guaranteed regardless of orientation [Tran 2013, Xuan 2004]. The high lifetime and stability of heat pipes during mechanical vibrations make heat pipes attractive for automotive applications [Tran 2013, Connors 2009]. Wang et. al. [2014] stored heat pipes using water as the working fluid in cold (-20°C) conditions for 14 hours, and functionality was immediately restored when heated.

In terms of battery thermal management, the low temperature gradient across the heat pipe facilitates an even temperature distribution. The high effective conductivity allows for heat to be transferred away from the battery to a remote heat exchanger where more space may be available. Recently, the amount of research considering heat pipe use for battery thermal management has grown. Constant temperatures below 40°C and increases in heat transfer of over 20 to 30% versus a solid material fin have frequently been reported in the literature [Tran 2013, Wang 2014, Belaev 2014, Mahefkey 1971, Rao 2012, Wu 2002]. Patents in the field are also becoming more prevalent [Huehner 2012, Walser 2014]. Most research has focused on individual heat pipe performance instead of the effect of such a system on the actual battery cells. Therefore, the focus of this research is the effect of a heat pipe BTMS on the battery cells, which are modeled using the SBC.

The following section describes briefly the functionality and production of a heat pipe. Section 7.2 presents the multifactorial analysis of the three layouts considered. Recommendations for the design of a BTMS for a luxury, high-performance PHEV are made in Section 7.3.

7.1. Layout and Considered Production Techniques

A heat pipe functions based on the same basic mechanism as a thermosyphon: heat exchange driven by the phase change of water and no external power sources [Dunn 1982]. In a thermosyphon, water vaporizes at the hot end of an evacuated and sealed tube, rises to the cold end of the heat pipe where it is condensed, and flows back via gravity to the hot side. The large latent heat of evaporation means that very high heat loads can be transferred over a very small temperature gradient; therefore, the effective conductivity is very high [Dunn 1982]. A heat pipe distinguishes itself from the thermosyphon in that it can operate independently of the gravitational force. In this research, the capillary force within a so-called "wick" along the inner wall facilitates circulation independent of the gravitational field. However, the term heat pipe also includes devices relying on centrifugal forces, osmosis, magnetism or electrohydrodynamics to return liquid water to the evaporator [Dunn 1982].

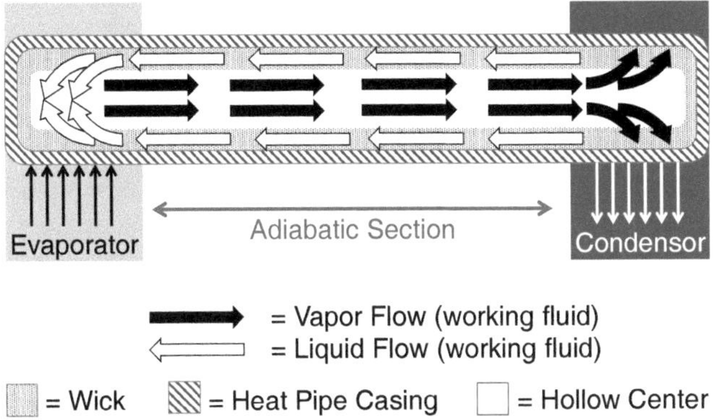

Figure 7.1: Basic heat pipe layout showing the flow and phase change of the working fluid, as well as the heat input at the evaporator (left) and transfer to the condenser (right).

The heat pipe is driven by the capillary head (Δp_c), which must exceed the sum of the pressure drop in the liquid phase from condenser to evaporator (Δp_l), the pressure drop in the vapor flow from condenser to evaporator (Δp_v), and the gravitational head (Δp_g), as shown in Equation 7.1 [Dunn 1982].

$$(\Delta p_c)_{max} > \Delta p_l + \Delta p_v + \Delta p_g \tag{7.1}$$

These various factors are influenced by the design and material of the wick, which must be tailored to the specific application of the heat pipe. For example, a fine powder wick with a small pore size increases the maximum capillary pressure, but at the price of a higher pressure drop for the liquid. The resulting lower velocity increases the risk of dry out in the evaporator, and therefore a loss in heat pipe functionality. A fiber wick on the other hand facilitates good fluid flow over longer distances, but the potential for trapping air during the production process is greater; thus, the production complexity and costs are increased due to the higher temperature vacuum required for material outgassing [Reay 2014b]. Copper, nickel and aluminum the most common wick materials for the operating range considered materials; however, it is critical that the wick material be selected in conjunction with the working fluid and casing material to maximize heat pipe lifetime by minimizing corrosion [Dunn 1982, Reay 2014b].

The selection of the working fluid also determines the operational temperature range. The extreme limits of the operating temperature are set by the triple- and critical-point temperatures of the working fluid; the operating temperature range is generally reduced in order to avoid high internal pressures and low heats of vaporization near the critical point [Cengal 2007]. For low temperature applications such as batteries and most consumer electronics, water, acetone, alcohol, and similar fluids are utilized [Dunn 1982, Tran 2013, Wang 2014, Reay 2014b]. So-called nanofluids are also frequently mentioned in the literature to enhance the thermal conductivity of the working fluid [Wan

2014]. Models have shown that the most critical characteristic for the working fluid is its thermal conductivity [Arab 2014].

Material choices for the heat pipe casing include copper [Tran 2013, Wang 2014], aluminum [Ameli 2012, Chen 2012], stainless steel [Belyaev 2014] and even ceramic [Meisel 2014]. Casing material choice is also a function of wick material and the working fluid [Dunn 1982, Reay 2014a, Reay 2014b]. As the quantity of heat pipes in a battery system increases, the material cost and weight become more relevant: this must certainly be considered when analyzing a BTMS concept, as new materials can reduce heat pipe weight by as much as 80% [Yang 2011, Reay 2014b].

The critical steps in the production of heat pipes are the preparation of the casing, wick and working fluid, as well as testing the leak tightness. The casing and wick must be thoroughly cleaned to prevent contamination and ensure complete wetting, and outgassed so that vapor chamber is not filled with gasses released form the metals under the vacuum created [Cengal 2007, Reay 2014b]. Similarly, the working fluid must be purified, and ideally freeze-degassed in order to remove any dissolved gasses that could collect in the condenser during operation [Reay 2014b]. Testing the leak tightness ensures that the heat pipe can function properly, and is thus a critical step in heat pipe production.

Currently, most heat pipes are assembled by attaching the wick material to the casing, then sealing the ends of the heat pipe. Round heat pipes are then bent and flattened to create smaller heat pipes such as those used in consumer electronics. The bending of such heat pipes results in folds and inhomogeneous surfaces within the casing, potentially causing the wick and casing to separate. With the increasing prominence of additive production techniques (3D-printing / Laser Sintering), the production of heat pipes having complex shapes and small sizes will become more economical [Reay 2014b].

Various heat pipe layouts exist in addition to a traditional cylindrical heat pipe with two closed ends: prominent examples include flat plate heat pipes (FPHP) or vapor chambers [Tran 2013, Xuan 2004] as well as loop heat pipes (LHP) or pulsating heat pipes (PHP) [Groll 2002, Park 2009, Qu 2006]. For this study, a traditional cylindrical heat pipe with a copper casing, a nickel fiber wick, and water as the working fluid is flattened. This design was recommended by the manufacturer due to the minimum achievable thickness (down to 0.4 mm with the current production technology) and the quick transition from cooling to heating (the evaporator and condenser side can change within seconds). A homogeneous temperature distribution is simultaneously maintained across the contact surface to the battery cell. Copper/water heat pipes are commonly produced for the consumer electronics industry, thus the operating temperature is similar. The current experience and widespread implementation of such heat pipes makes them ideal from a production-oriented point of view.

The focus of this research is not the layout or make-up of the individual heat pipe, but rather the investigation of heat pipes in general for the purpose of enhancing cell performance. A technology demonstrator, presented in the following section, was designed to test multiple features and potential uses of heat pipes simultaneously.

For the technology demonstrator, flattened heat pipes with a final height of 1.2 mm are imbedded in and soldered to a 1.5 mm copper plate. Four heat pipes are evenly spaced across the plate. To conform to the geometry of the reference module and bypass the threaded rod, the heat pipes are bent. It should be noted that this shape would be undesirable in a mass-production due to the additional production time and the reduced performance of the heat pipe. However, to maintain the consistency of the comparison, the reference module geometry must be used.

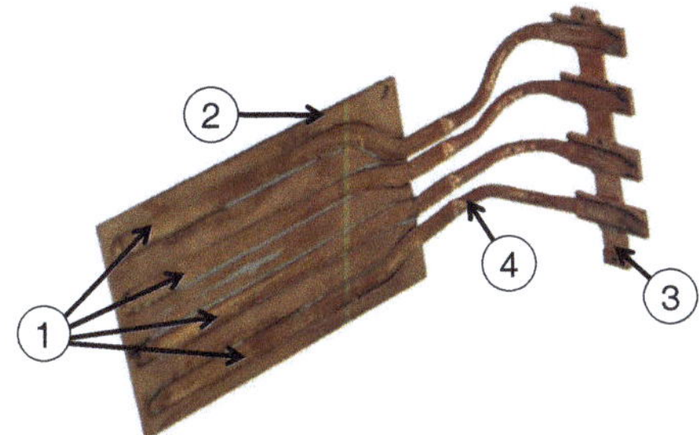

Figure 7.2: A heat-pipe spacer utilized for the feasibility study consisting of four flat heat pipes (1) imbedded in a copper plate (2). The complicated bends (4) proceeding to the interface plate (3) are a result of the geometry of the reference module.

The spacers interlock at the interface plate, which is designed to distribute the heat before further transport by the long heat pipes. The long heat pipes transfer heat to the remote cold plate, where the coolant is provided. This design separates the liquid coolant form the cells in order to prevent a large coolant leakage directly onto the cells in the event of a crash or other system failure.

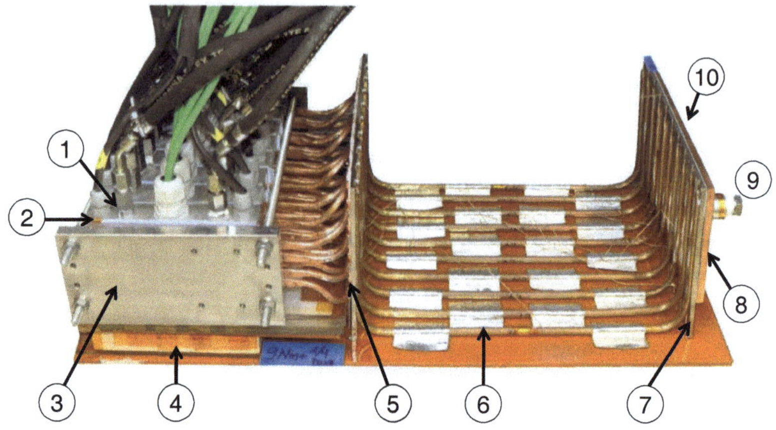

Figure 7.3: The complete heat-pipe technology demonstrator used for the feasibility study. The SBCs (1) are between heat pipe spacers (2) and compressed with the same compression plates and bolts (3). The insulating base (4) is elevated below the cells and continues under the entire set-up. The heat pipe spacers (2) are fastened to the interface plate (5) on one side via a thermal gap pad, while the long heat pipes (6) are soldered to the other side. The long heat pipes (6) connect to the remote cold plate (8) over a further interface plate (7). The inlet (9) and outlet (10) to the cold plate are marked as well.

The copper heat pipes used in the spacer each have a thermal resistance of approximately 2 K^1W^{-1}, and transfer approximately 10 W at an ambient temperature of 25°C. When a temperature below 40°C is maintained at the condenser, the temperature difference between the condenser and evaporator is 2 to 3°C. The temperature difference along the condenser or evaporator itself (i.e. the cell contact surface) is much smaller.

The 300 mm long heat pipes have greater temperature differences across their length. The contact between the heat pipe spacers and the interface plate is achieved using screws and a thermal gap pad to ensure even contact. This modular design was chosen so that spacers could be added and removed: a better thermal contact is certainly possible using an optimized design.

The setup itself is designed to test two principles simultaneously: first, the transfer of heat to or from the cells, and second the transfer of heat between the module and a remote heat exchanger. The first principle could be applied alone, with heat pipe spacers connected directly to a heat exchanger instead of the current interface plate. The second principle is designed to test the ability of heat pipes to transfer heat to a remote location, where more space is available and the majority of the coolant s further from the cells. The principles are tested together in one technology demonstrator to reduce feasibility study costs.

Because of the sub-optimal bends to adapt to the reference geometry, and the sub-optimal contact between heat pipe spacer and interface plate to allow for experimental versatility, the experimental results present a conservative indication of potential heat pipe performance.

7.2. Multifactorial Analysis of Heat-Pipe Spacers

For this layout, the number of spacers was varied: one heat pipe spacer between every cell, one every second cell, and one every fourth cell. The amount of long heat pipes between the distributer plate and the cold plate / heat exchanger were held constant. The thermal performance and theoretical pump power are experimentally measured at various ambient temperatures, coolant flow rates, and coolant inlet temperatures using a 50/50 water-glycerol. In all cases, the boundary conditions are held constant throughout the trial. To facilitate clarity in the plots throughout the research, the case of 20°C coolant inlet temperature and 5 L^1min^{-1} flow rate is shown. As expected, a lower coolant inlet temperature and higher flow rate decreases module average temperature. This trend is consistent over all trials. The multifactorial analysis is presented below.

Thermal Performance

The thermal performance of the technology demonstrators is evaluated based on the criteria presented in Section 2.1, and plotted for the three cases analyzed in Figures 7.4 through 7.6.

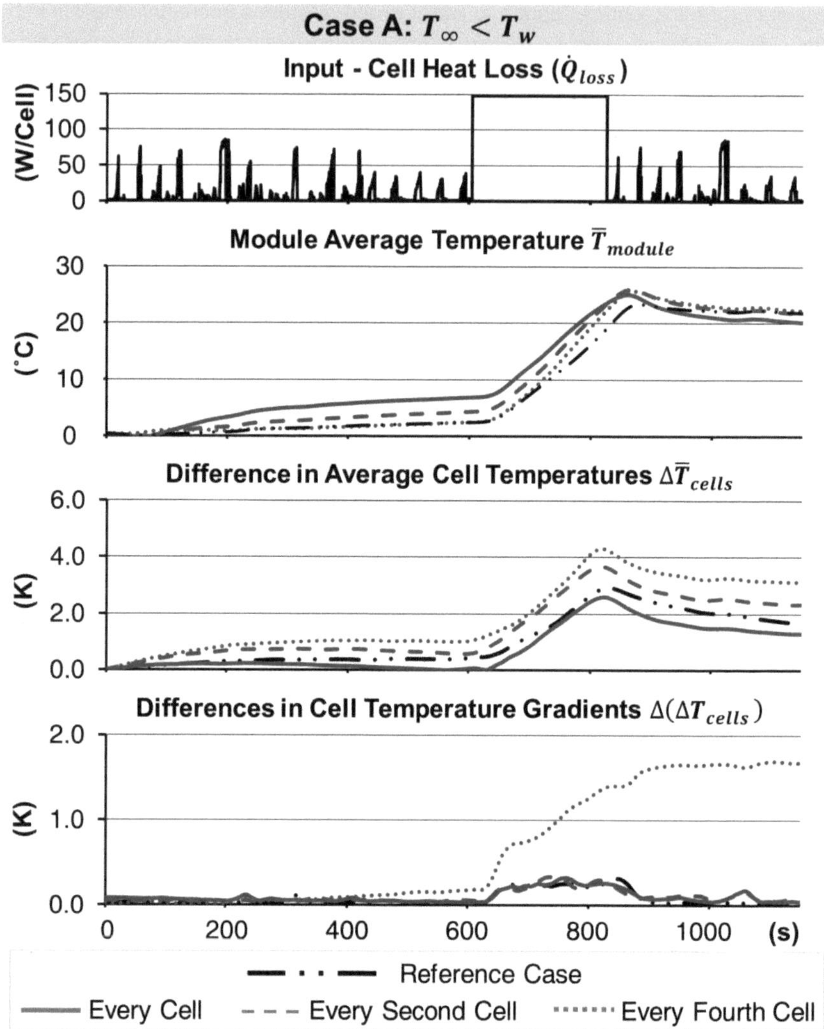

Figure 7.4: The cell het loss profile (input) for case A (T_∞=0°C and T_w=20°C) is shown at top, followed by the three criteria for thermal performance (Section 2.1):, the average module temperature $\bar{T}_{module}$, the difference in average cell temperature $\Delta\bar{T}_{cells}$, and the differences in the magnitude of the cell temperature gradients $\Delta(\Delta T_{cells})$. Data at a constant coolant flow rate (5 L^1min^{-1}) and inlet temperature (20°C) is shown. All plots share the same x-axis.

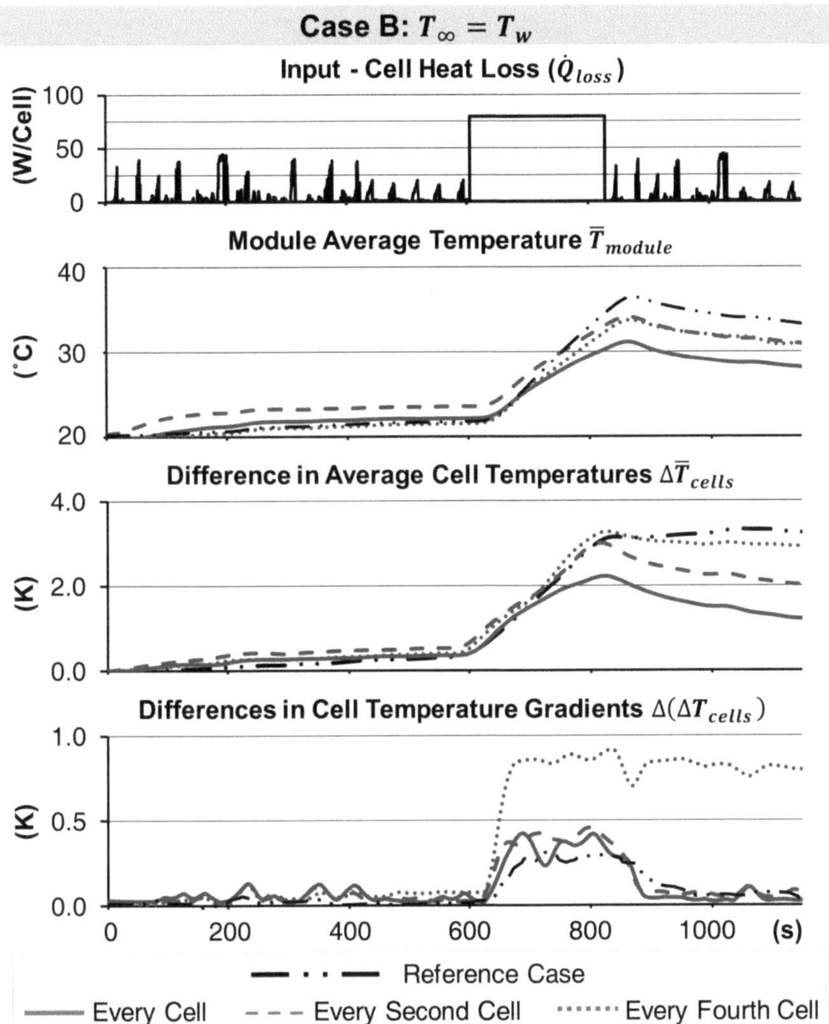

Figure 7.5: The cell het loss profile (input) for case B (T_∞=20°C and T_w=20°C) is shown at top, followed by the three criteria for thermal performance (Section 2.1):, the average module temperature $\bar{T}_{module}$, the difference in average cell temperature $\Delta\bar{T}_{cells}$, and the differences in the magnitude of the cell temperature gradients $\Delta(\Delta T_{cells})$. Data at a constant coolant flow rate (5 L^1min^{-1}) and inlet temperature (20°C) is shown. All plots share the same x-axis.

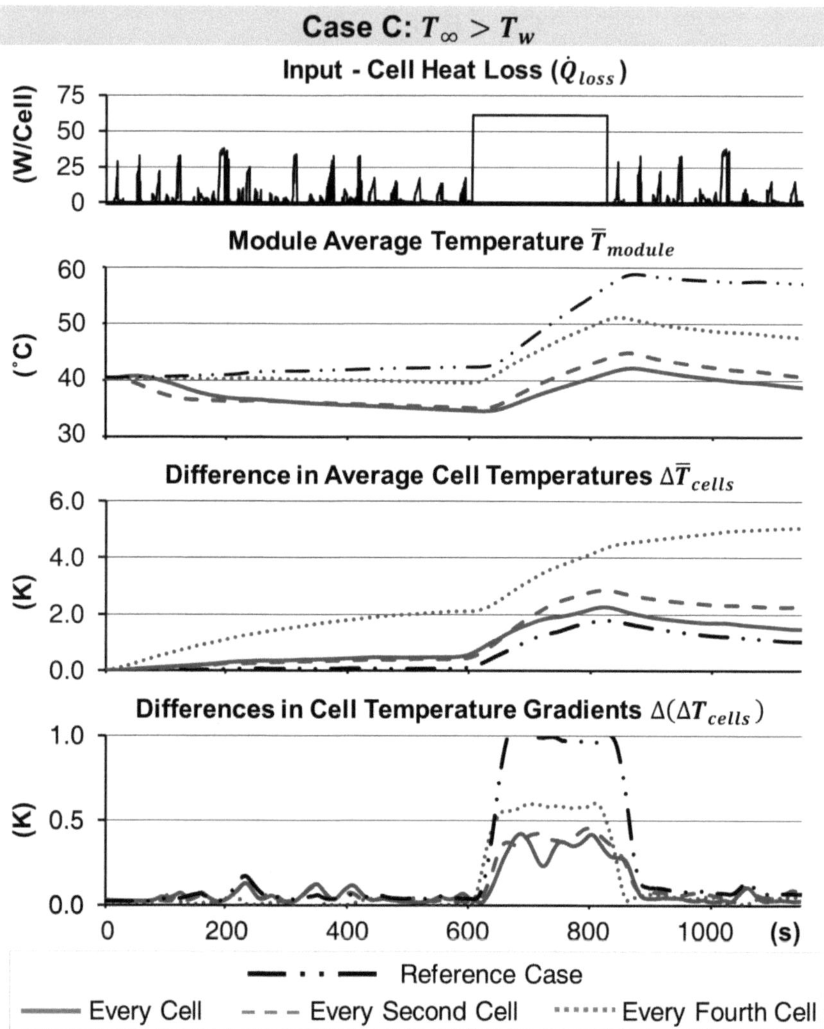

Figure 7.6: The cell het loss profile (input) for case C (T_∞=0°C and T_w=20°C) is shown at top, followed by the three criteria for thermal performance (Section 2.1):, the average module temperature $\bar{T}_{module}$, the difference in average cell temperature $\Delta\bar{T}_{cells}$, and the differences in the magnitude of the cell temperature gradients $\Delta(\Delta T_{cells})$. Data at a constant coolant flow rate (5 L¹min⁻¹) and inlet temperature (20°C) is shown. All plots share the same x-axis.

As for the reference case and other technology demonstrators, the solution to the first law of thermodynamics is approximated for the average module temperature (Equation

5.2). The fitted coefficients for the thermal conductance to/from the ambient and the BTMS are shown in Table 7.1.

Table 7.1: Iteratively calculated values for the heat capacity (C), the thermal conductance to/from ambient (hA_∞), and the thermal conductance to/from the BTMS (hA_{BTMS}) at the three cases analyzed (described in Section 4). The goodness of fit for of the numerical solution is also included (r^2).

	Case	C (J^{1}K^{-1})	hA_∞ (W^{1}K^{-1})	hA_{BTMS} (W^{1}K^{-1})	Fit (r^2)
	A	9600	5.4	4.5	0.999
Every Cell	B	9600	5.3	8.1	0.999
	C	9600	5.4	8.8	0.995
	A	9200	5.3	1.8	0.999
Every Second Cell	B	9200	5.3	4.2	0.998
	C	9200	5.4	6.1	0.994
	A	9000	5.3	0.8	0.999
Every Fourth Cell	B	9000	5.3	2.8	0.999
	C	9000	5.3	4.0	0.996

The heat capacity is rounded to the nearest hundred, and the starting value for the iterative fitting process was the calculated value. The high heat capacity of the heat pipe technology demonstrator in comparison to the other concepts analyzed is in part due to the remote heat pipes and two spreader plates.

The values of thermal conductance to the ambient determined for the reference case were used as initial guesses for fitting the solution; however, due to the large surface area of the BTMS that is exposed to the ambient (A_{BTMS}), the conductance is higher than in other cases. The boundary condition (natural convection, constant ambient temperature) is the same for each case (A through C) across the technology demonstrators.

In contrast to other concepts, the thermal conductance of the BTMS is not constant across cases A through C. This indicates a temperature dependence of the effective thermal conductance over the heat pipe, which can be seen in Figures 7.4 through 7.6 in the behavior of the average module temperature. As a result, the thermal conductance increases from case A to B to C, because the ambient (and initial) temperature is increased from 0°c to 20°C to 40°C. Because of the heat pipe makeup (copper-water), such temperature dependence is expected. This is due to the working fluid, water, which approaches its freezing point in case A, and therefore has an increased viscosity, which in turn increases the pressure drop over the heat pipe, limiting its effectiveness.

The evaluation criteria pertaining to the spatial temperature variations are also influenced by temperature gradients in the spreader plates and the remote heat pipes. Thus, the gradients shown reflect a very conservative (worst-case) estimate of $\Delta \bar{T}_{cells}$ and $\Delta(\Delta T_{cells})$ with respect to an optimized (only heat pipe spacers) layout. However, for this study the flexibility of the technology demonstrator (i.e. adding and removing spacers) was prioritized over a single optimal layout. The functionality (and thermal conductance to the BTMS) of the demonstrator would most likely be better for heat pipe spacers alone, without the additional thermal resistance created by the junctions to the spreader plates and over the 300 mm remote heat pipes.

The lack of direct contact (i.e. a spacer every fourth cell) between a cell and the BTMS causes diverse gradients over the individual cells, indicating the potential for inhomogeneous aging. The configuration of a heat pipe spacer between every cell and every second cell results in both low temperature differences between the cells ($\Delta \bar{T}_{cells}$) and small variations in the magnitude over the cells ($\Delta(\Delta T_{cells})$), though again the values of $\Delta(\Delta T_{cells})$ fall primarily within the measurement uncertainty of the thermocouples (Section 3).

Considering the multiple thermal interfaces along the heat transfer path from the cell to the heat pipe spacer and finally to the cold plate, the total thermal performance of the demonstrator is exceptional, demonstrating the potential of heat pipes. This concept is compared to the other thermal management concepts in Section 8.

Hydraulic Work

The heat exchanger used in the technology demonstrator consisted of two standard cold plates ($\approx$100 x 100 mm), and was not altered during the experimental trials. Thus, theoretical pumping power is nearly constant across all cases at a specified flow rate and coolant temperature. For this reason, the hydraulic work of the technology demonstrator is not analyzed further. Rather, it is assumed that any flow-optimized heat exchanger can be utilized in conjunction with heat pipe spacers and/or remote heat pipes.

Vehicle Suitability

The vehicle suitability (Section 2.3) of the heat pipe spacer layouts in comparison to one another is shown for both the existing technology demonstrator and for potential future (optimized) layouts in Table 6.2. The "current" values are directly measured, while the potential layouts are in this case based on flat heat pipes embedded in a 1 mm aluminum plate, as shown in Figure 7.7. In this way, weight savings versus the copper plates in the technology demonstrator are realized.

Table 7.2: Multifactorial analysis of the heat pipe spacer technology demonstrators based on the objective evaluation criteria (Section 2.1). The values shown are the average over the cases (A through C) tested.

	Spacer between each Cell		Spacer each Second Cell		Spacer each Fourth Cell	
Thermal Behavior						
Thermal Conductance [W^1K^{-1}]	6.2		4.0		2.5	
Heat Capacity [J^1K^{-1}]	9600		9200		9000	
Maximum $\overline{\Delta T}_{cells}$ [K]	2.6		3.7		5.0	
Maximum ΔT_{cell} [K]	1.8		1.9		1.9	
Maximum $\Delta(\Delta T_{cells})$ [K]	0.4		0.4		1.7	
Hydraulic Work						
Theoretical Pump Power[1] [W]	10.1		10.1		10.1	
Vehicle Suitability[2]	*Current*	*Potential*	*Current*	*Potential*	*Current*	*Potential*
Weight: Δm	+ ≈ 28%	+ ≈ 4%[3]	+ ≈ 16%	+ ≈ 2%[3]	+ ≈ 9%	+ ≈ 1%[3]
Volume: ΔV	+ ≈ 6%	+ ≈ 4%[3]	+ ≈ 3%	+ ≈ 2%[3]	+ ≈ 2%	+ ≈ 1%[3]
Length: ΔL_x	+ ≈ 6%	+ ≈ 4%	+ ≈ 3%	+ ≈ 2%	+ ≈ 2%	+ ≈ 1%
Width: ΔL_y	+ ≈14%[4]	+ ≈ 6%[4]	+ ≈14%[4]	+ ≈ 6%[4]	+ ≈14%[4]	+ ≈ 6%[4]
Heigth: ΔL_z	+ ≈ 0%	+ ≈ 0%	+ ≈ 0%	+ ≈ 0%	+ ≈ 0%	+ ≈ 0%

[1] two basic cold plates connected in series at a flow rate of 5 L^1min^{-1}, averaged over all trials
[2] in comparison to an 8-cell module with no thermal management, excluding the mechanical bracing
[3] using a 1 mm aluminum plate with 0.8 mm embedded heat pipes
[4] including the bending radius to the interface plate

Producibility and Economic Viability

In terms of the production complexity, the copper/water heat pipe is widely used in the consumer electronics industry. They can be produced in various sizes, with thickness of 0.4 mm possible. At this size, however, the heat transfer per heat pipe decreases, so that the number of heat pipes required increases. On the other hand, larger diameter heat pipes are easier to produce and can transfer more heat, but the module length can increase significantly when integrated in a spacer configuration. The reduced production complexity and cost utilizing the current production methods speak for heat pipe of a least 1.5 to 3 mm thickness. At this size, the use of a spacer-based BTMS however increases the volume of the module significantly, making a cooling plate layout more attractive for vehicles used under average driving conditions. However, the implementation of additive techniques presents an opportunity to realize smaller, more complex geometries and wick structures. In terms of the life cycle energy consumption, copper is superior to primary aluminum in the material production; however, this benefit must be balanced against the higher weight and therefore higher energy consumption during the use of the vehicle.

Summary of the Multifactorial Analysis

Heat pipes present a high potential for battery thermal management due to their widespread use in other related sectors. The technology demonstrator considered in this research seeks to explore the primary benefit of heat pipes in terms of safety: very little fluid is directly near the battery system. The remote heat pipes show the potential of transferring heat over a greater distance with not energy input. The general conclusions from the multifactorial analysis are summarized as follows:

- Thermal contact to the BTMS must be realized for all cells (or none): the temperature gradients created by contact with only certain cells increase the risk of inhomogeneous aging versus the reference case;
- The thermal resistance due to the two spreader plates and over the remote heat pipes decreases the total thermal conductance of the demonstrator;
- Water-copper heat pipes are least effective at low temperatures; however, a rapid reaction time was observed once the temperature increased from 0°C indicating the potential of the technology using an optimized layout;
- Standard heat pipes can already be produced cost effectively. New heat pipe layouts and complex geometries, especially using additive production techniques, will make heat pipes even more interesting from both a performance and production/cost standpoint.

Based on the conclusions from the multifactorial analysis, recommendations are made for the application of heat pipes for battery thermal management in a high-performance PHEV.

7.3. Discussion of Results and Design Recommendations

It has been shown by this research that heat pipes provide a promising method for battery thermal management. Heat pipe thermal management for battery systems can additionally be influenced by reducing the optimal operating range of the heat pipes (material combination), or increasing the heat pipe size and/or quantity. Increasing the size and/or quantity of heat pipes compensates for the inefficient operation of copper/water heat pipes at lower temperatures.

The results also indicate that the thermal resistance over the remote heat pipes is significant (reflected by the high temperature differences from condenser to evaporator of 5 to 15°C). It is therefore recommended that both alternative heat pipe material combinations and layouts be analyzed based on the evaluation criteria presented in this research (Section 2). Other heat pipe forms (FPHP, LHP, PHP, etc.) and material combinations (casing, wick, working fluid) may improve performance. Research on material combinations with the goal of lowering the weight of individual heat pipes is critical for battery system thermal management, due to the large quantities of heat pipes that would be required in a large battery system. To further enhance safety, the working fluids considered should be tested on their electrical conductivity, as in the event of a crash, the fluid could leak onto the battery.

Based on the high thermal resistance of the remote heat pipes utilized, and because of the added cost and weight in this dimension, the optimization of shorter heat pipes to transfer heat to/from the cells is recommended as a first step.

Recommendation for a High-Performance PHEV

The technology demonstrator, with its copper plate and complex bends, does not represent the optimal layout in terms of vehicle suitability. To demonstrate the feasibility of a spacer suitable for vehicle integration, a 1 mm plate with three embedded plat heat pipes was produced and is shown in Figure 7.7. The weight and size reductions exhibit the potential of the technology. Such a spacer could provide adequate heat transfer when combined with a heat exchanger at the side of the module.

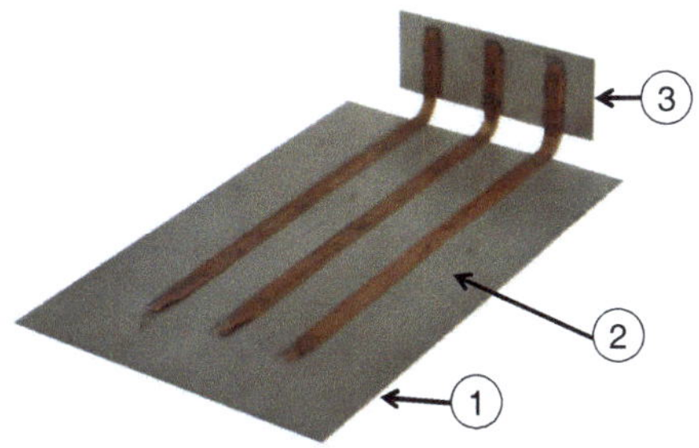

Figure 7.7: An optimized spacer to demonstrate the potential of a heat pipe spacer for a vehicle-suited application. Shown are a 1 mm aluminum plate (1), a flattened 0.8 mm copper / water heat pipe (2) that can be connected to a heat exchanger at the opposite end (3).

The production complexity of these thin and long heat pipes is currently high, while the heat transfer is limited. By including larger diameter heat pipes, heat removal is guaranteed even at inefficient (cold) operating temperatures. As a result, heat pipe length could be increased, allowing for the placement of a heat exchanger at an ideal location in terms of the vehicle suitability. Considering not only the technical benefits but also the decreased production complexity and cost of larger (1.5 to 3mm) copper/water heat pipes, a spacer may not be optimal in terms of vehicle suitability: for PHEV-2 cells, a cooling plate consisting of multiple heat pipes embedded in an aluminum plate would provide adequate heat removal in a smaller packaging (< 3mm height) than would be possible for a cooling plate using a liquid-coolant because of the resulting pressure drop. Because more, larger heat pipes are used, the distance to the heat exchanger could be increased. The heat pipe itself utilizes no external power source, meaning that the energy consumption of the BTMS is a function of the heat exchanger used. Such a heat exchanger could then be optimized for pressure drop, and placed away from the battery system for enhanced safety in a crash and/or to take advantage of available space elsewhere. A schematic of this concept is shown in Figure 7.8.

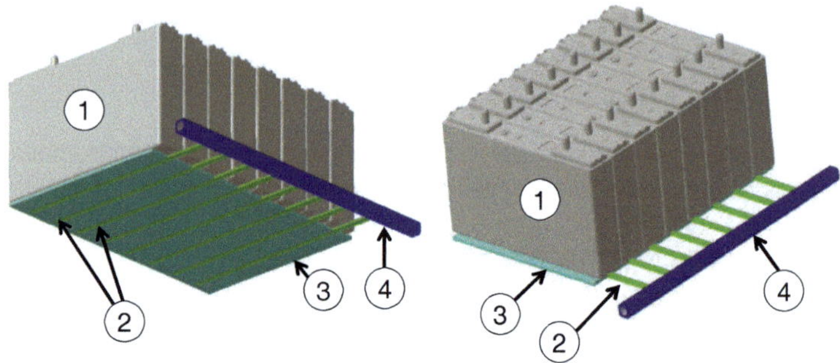

Figure 7.8: The recommended implementation of heat pipes for the thermal management of PHEV-2 format cells, showing a battery module (1) on a cooling plate made up of flattened heat pipes (2) embedded in an aluminum plate (3). The opposing ends of the heat pipe are thermal contacted to an heat exchanger optimized for pressure drop.

Heat pipes show great potential. The technology demonstrator shows a very conservative estimate of the performance due to the sub-optimal heat pipe bends and contact at the interface plates designed to accommodate the reference module considered in this research. These compromises were however necessary to make a comparison with other concepts possible. The true benefit of long heat pipes would be the ability to use a larger, cheaper heat exchanger, possibly even with external or cabin air, at a location in the vehicle where more space is available. Such a layout could result in a thermal management system that uses no additional energy, thereby enhancing vehicle range.

The heat pipe technology in its current state is compared against other promising BTMS concepts in Section 8. The potential of the technology observed is considered in the comparison.

8.   Inter-Concept Comparison

In the previous sections, variations of individual thermal management concepts are compared to one another in order to identify critical design features, and to assist in optimizing the BTMS for an automotive application. This section serves to compare the concepts amongst one another. This global comparison between completely different concepts is possible due to the novel methodology presented in Sections 2 and 3. Because only the BTMS was changed between trials (and not the cells, contact surface, thermocouple position, etc.), this comparison isolates the effect of the BTMS alone on the cells. In this section, the most promising configurations identified in Sections 5 through 7 are compared to one another: cooling plate technology demonstrator 1, a ceramic spacer every second cell, and a heat pipe spacer every second cell connected to a cold plate via remote heat pipes.

The analysis of the reference module without thermal management (Section 4) demonstrated that around an ambient temperature of 20°C (Case B), thermal management is only required at high, sustained heat loads. Depending on the region where the vehicle is used, such temperatures are more or less frequent; however, predicting where a vehicle will ultimately operate is difficult, and adverse temperatures will most likely be encountered. For this reason, the use cases of a cold- and hot-soak (0°C and 40°C ambient temperature, respectively) are considered in order to analyze the performance of the BTMS at extreme operating conditions.

8.1.   Use Case 1: Hot-soak at 40°C

The passenger cabin of a vehicle standing in the sun throughout the day during the summer months reaches very high temperatures. The same occurs for the battery system. This use case considers a battery system that has homogeneously reached a temperature of 40°C, and must then operate in the summer heat. In this case, the implementation of a BTMS allows for heat to be removed from the battery system, whereas a battery system without thermal management has little chance of dissipating heat.

The thermal performance of the most promising BTMS concepts analyzed in this research are compared in Figure 8.1. For the ceramic spacer, a flow rate of 5 L^1min^{-1} results in a significantly higher theoretical required pumping power (over 6 W versus under 1 W for the cooling plate). This increased energy consumption is partly, but not entirely, due to the sub-optimal inlet and outlet design of the technology demonstrator. Therefore, the effect of reducing the coolant flow rate to 2 L^1min^{-1} is shown by the dashed line. The result of this reduction in flow rate is a theoretical pumping power of under 0.5 W. Thus, the energy consumption is halved versus the cooling plate, while thermal performance is increased.

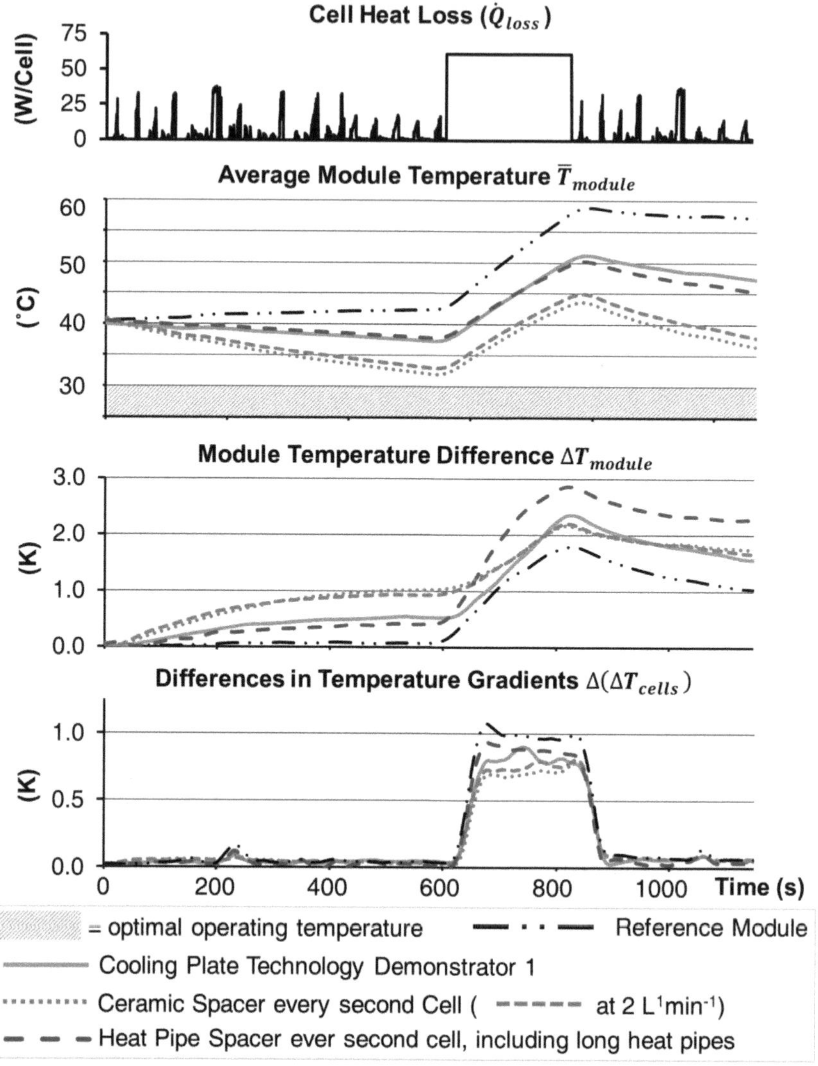

Figure 8.1: The cell heat loss profile (input) is shown at top, followed by the three criteria for thermal performance (Section 2.1): the average module temperature $\bar{T}_{module}$, the average cell temperature differences $\Delta\bar{T}_{cells}$, and the differences in the magnitude of the cell temperature gradients $\Delta(\Delta T_{cells})$ at 40°C ambient temperature (Case C, Section 4) for the most promising BTMS from Sections 5 through 7. The results are shown at a constant inlet temperature (20°C) and constant coolant flow rate of 5 L^{1}min^{-1} (all demonstrators) and 2 L^{1}min^{-1} (ceramic spacer). All plots share the same x-axis.

A ceramic spacer between every second cell increases the area of the thermal contact surface of the BTMS (A_{BTMS}) by nearly three-and-a-half times versus a cooling plate under the module. The effect of the increased surface area while using the same coolant flow rate and inlet temperature is shown in Figure 8.1, as the ceramic spacer reduces module average temperature significantly over the cooling plate. The slope of the average temperature curve is steeper than that for the cooling plate or the heat pipe technology demonstrator, indicating that the spacer layout reacts more quickly after a hot-soak. The differences in average module temperature are quantified by the fitted coefficients to the first law of thermodynamics from Sections 5 through 7 (Table 8.1).

Table 8.1: Iteratively calculated values for the heat capacity (C), the thermal conductance to/from ambient (hA_∞), and the thermal conductance to/from the BTMS (hA_{BTMS}) at 40°C ambient temperature (Case C, Section 4). The goodness of fit for of the numerical solution is also included (r^2).

	C (J^{1}K^{-1})	hA_∞ (W^{1}K^{-1})	hA_{BTMS} (W^{1}K^{-1})	Fit (r^2)
Reference Case (No Thermal management)	6500	3.5	-	0.998
Cooling Plate Demonstrator 1	7200	3.0	6.2	0.994
Ceramic Spacer every second cell (5 L^{1}min^{-1})	7100	3.1	10.8	0.995
Ceramic Spacer every second cell (2 L^{1}min^{-1})	7100	3.2	10.5	0.995
Heat Pipe Spacer every second cell	9200	5.4	6.1	0.994

After a 40°C hot-soak, the experimentally analyzed BTMS improve the average temperature of the module $\bar{T}_{module}$ and reduce the differences in the temperature gradients between cells $\Delta(\Delta T_{cells})$. However, as expected, gradients are induced across the module ($\Delta \bar{T}_{cells}$) as the result of the increased heat transfer at the BTMS surface; however, only at high heat loads does the value of $\Delta \bar{T}_{cells}$ exceed the measurement uncertainty of the thermocouples (Section 3). The high value of thermal conductance to the ambient (hA_∞) for the heat pipe technology demonstrator is due to the larger area exposed to the ambient (Section 7).

The ceramic spacers initially induce higher temperature differences amongst the cells due to the high thermal conductance. The cooling plate and the heat pipe spacer connected to the cold plate over remote heat pipes demonstrate similar behavior; a heat pipe spacer alone would certainly exceed the performance of the cooling plate. This data demonstrates the potential of heat pipes in battery thermal management, especially at higher temperatures where the copper-water heat pipes used operate more efficiently.

If constant use at a high cell heat loss is anticipated, the streamed spacer provides the highest heat transfer to maintain the ideal operating range; however, such conditions would most likely only occur in frequent track driving and in certain adverse climates. A cooling plate connected at the foot of the prismatic cells improves the thermal performance of the module versus the reference case, and provides the opportunity to pre-condition a battery system before use. In terms of the temperature gradients across the module, the order of magnitude of the concepts presented in this Section is similar.

8.2. Use Case 2: Cold-start at 0°C

Battery performance is very limited at 0°C: high rates of discharging and especially charging are detrimental to battery system health (Section 1). A cold start at or below 0°C is however very common in the winter months when a vehicle is left standing outside. Two effects contribute to reduced range and performance at this point: first, the cells are far below their ideal operating range, resulting in a higher internal resistance and thus lower efficiency. Second, the climatization of the passenger cabin requires energy, which must come from the battery in the case of a BEV: a PHEV or BEV with a range extender can take advantage of the internal combustion engine to reduce the load on the battery. An efficient BTMS thus allows for the battery to be used sooner, as well as for the pre-conditioning of the vehicle cabin and battery system while the vehicle is charging.

The thermal performance of the most promising BTMS concepts analyzed in this research are compared in Figure 8.2. The results for the ceramic spacer technology demonstrator using a lower flow rate (2 L^1min^{-1}) are also shown: even at half the theoretical pumping power, the spacer provides greater heat transfer than the state of the art, a cooling plate.

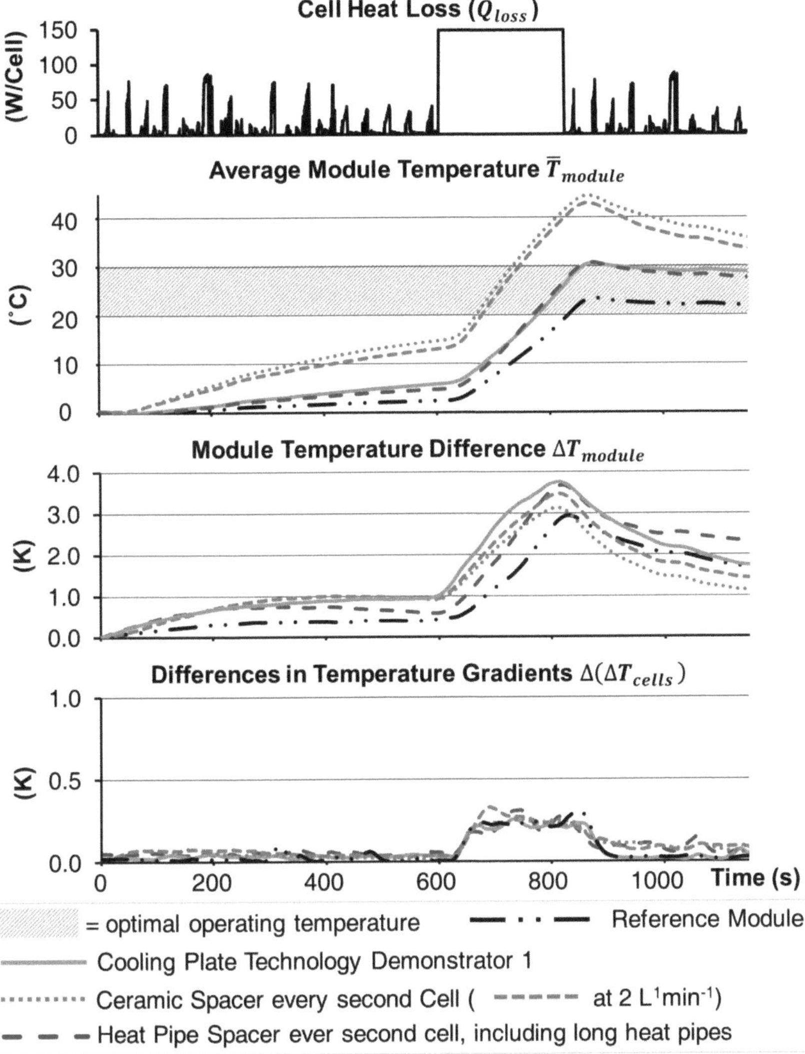

Figure 8.2: The cell heat loss profile (input) is shown at top, followed by the three criteria for thermal performance (Section 2.1): the average module temperature $\bar{T}_{module}$, the average cell temperature differences $\Delta\bar{T}_{cells}$, and the differences in the magnitude of the cell temperature gradients $\Delta(\Delta T_{cells})$ at 0°C ambient temperature (Case A, Section 4) for the most promising BTMS from Sections 5 through 7. The results are shown at a constant inlet temperature (20°C) and constant coolant flow rate of 5 L^{1}min^{-1} (all demonstrators) and 2 L^{1}min^{-1} (ceramic spacer). All plots share the same x-axis.

The average module temperature during the driving cycle after a cold start again shows the superior heat transfer of the water-glycol streamed spacer between every second cell (Figure 8.2). This behavior is again quantified by the fitted coefficients to the first law of thermodynamics from Sections 5 through 7 (Table 8.2).

The order of magnitude of the spatial temperature variations are similar across the BTMS analyzed; only the values of $\Delta \bar{T}_{cells}$ at high heat loads exceed the measurement uncertainty of the thermocouples in the SBC (Section 3). As expected, the induced gradients are larger than in the reference case, where the module transfers heat only by natural convection to the ambient. Due to the y-axis scale of the plot of average module temperature (Figure 8.2), the heat pipe technology demonstrator appears to perform similarly to the cooling plate; however, when examining the difference more closely, the coefficients for the fitted solution to the first law of thermodynamics (Table 8.2.) depict the behavior more accurately: the efficiency of heat pipe demonstrator near 0°C is low, where the heat transfer to the ambient is dominant.

Table 8.2: Iteratively calculated values for the thermal capacity (C), the thermal conductance to/from ambient (hA_∞), and the thermal conductance to/from the BTMS (hA_{BTMS}) at 0°C ambient temperature (Case A, Section 4). The goodness of fit for of the numerical solution is also included (r^2).

	C (J^{1}K^{-1})	hA_∞ (W^{1}K^{-1})	hA_{BTMS} (W^{1}K^{-1})	Fit (r^2)
Reference Case (No Thermal management)	6500	3.0	-	0.997
Cooling Plate Demonstrator 1	7200	3.0	6.2	0.996
Ceramic Spacer every second cell (5 L^{1}min^{-1})	7100	3.2	10.9	0.998
Ceramic Spacer every second cell (2 L^{1}min^{-1})	7100	3.2	10.5	0.998
Heat Pipe Spacer every second cell	9200	5.3	1.8	0.999

For the cold-start use case, all BTMS analyzed improve the average module temperature. The decreased efficiency of copper-water heat pipes is most likely amplified by the high thermal resistance of the remote heat pipes.

8.3. Multifactorial Analysis Across Both Use Cases

Because the remaining evaluation criteria are not affected by the ambient temperature, they are shown together in this section. The general thermal performance is also shown across use cases in this Section to provide a clear comparison of the BTMS concepts.

Not only thermal performance should drive the selection of the ideal BTMS for certain use cases: the energy consumption, size, production complexity and economic viability must also be considered. The energy consumption is shown in the representative case utilized throughout the plots shown in this research: 20°C coolant inlet temperature and a flow rate of 5 L^1min^{-1} (or 2 L^1min^{-1}). It should be noted that, as presented in Section 7, the heat pipe technology demonstrator utilizes two standard cold plates connected in series. In an optimized system, the remote heat exchange could be optimized for pressure drop, and the energy consumption of the total system would then be lower.

The order of magnitude of the temperature differences between the cells ΔT_{module} and the difference in the magnitude of the temperature gradients between the cells $\Delta(\Delta T_{cells})$ are higher at higher cell heat loss, but only outside the measurement uncertainty in the case of $\Delta \bar{T}_{cells}$. In all cases, the price of a lower average temperature is a temperature gradient across the cells/module.

Table 8.3: Multifactorial analysis of the reference case compared to cooling plate demonstrator 1, a ceramic spacer every second cell, and a heat pipe spacer every second cell. The data is drawn from the respective Section (5 through 7) and shown based on the objective evaluation criteria (Section 2.1). The values of thermal performance are combined across the use cases.

	Reference Case (No Thermal Management)	Cooling Plate Technology Demonstrator 1		Ceramic Spacer every Second Cell		Heat Pipe Spacer every Second Cell	
Thermal Behavior							
Thermal Conductance [W^1K^{-1}]	3.3	6.2		10.9 $(10.5)^2$		4.0	
Heat Capacity [J^1K^{-1}]	6500	7200		7100		9200	
Maximum $\Delta \bar{T}_{cells}$ [K]	2.9	3.8		3.4 $(3.3)^2$		3.7	
Maximum ΔT_{cell} [K]	1.4	1.7		1.8 $(1.9)^2$		1.9	
Maximum $\Delta(\Delta T_{cells})$ [K]	1.1	1.3		0.8 $(0.8)^2$		0.4	
Hydraulic Work							
Theoretical Pump Power[1] [W]	0	1.2		5.9 $(0.5)^2$		10.1	
Vehicle Suitability[3]	*Current*	*Current*	*Potential*	*Current*	*Potential*	*Current*	*Potential*
Weight: Δm	+ 0%	+ ≈ 12%	+ ≈ 5%[4]	+ ≈ 12%	+ ≈ 4%[5]	+ ≈ 16%	+ ≈ 2%[7]
Volume: ΔV	+ 0%	+ ≈ 9%	+ ≈ 9%[4]	+ ≈ 8%	+ ≈ 3%[5]	+ ≈ 3%	+ ≈ 2%[7]
Length: ΔL_x	+ 0%	+ ≈ 18%	+ ≈ 10%	+ ≈ 8%	+ ≈ 2%	+ ≈ 3%	+ ≈ 2%
Width: ΔL_y	+ 0%	+ ≈ 0%	+ ≈ 0%	+ ≈ 7%[6]	+ ≈ 3%[6]	+ ≈14%[8]	+ ≈ 6%[8]
Heigth: ΔL_z	+ 0%	+ ≈ 5%	+ ≈ 5%	+ ≈ 0%	+ ≈ 0%	+ ≈ 0%	+ ≈ 0%

[1] at a flow rate of 5 L^1min^{-1}, averaged over all trials
[2] at a flow rate of 2 L^1min^{-1}, averaged over all trials
[3] in comparison to an 8-cell module with no thermal management, excluding the mechanical bracing
[4] depending on the necessary structural layout and material combinations, full coverage of module base
[5] using aluminum extruded spacers with soldered or glued distributer at end
[6] including the optimization of the distributer channel routing
[7] using a 1 mm aluminum plate with 0.8 mm embedded heat pipes
[8] including the bending radius to the interface plate

The values in Table 8.3 are carried over from the respective Sections (5 through 7). The ceramic spacer every second cell was thermally superior, increasing the thermal conductance significantly while minimally increasing the temperature difference over individual cells and between the cells in comparison to the reference case. The minimum theoretical pump power was also be achieved with the ceramic spacers because the high thermal conductance facilitates the use of low fluid flow velocities. A heat pipe spacer attached directly to a heat exchanger (without the remote heat pipes and spreader plates) may provide a similar energy saving potential, as the heat pipes could be connected to a simple, pressure-drop optimized cooling plate. The spacer arrangements, as expected, result in a more homogeneous temperature distribution within the module versus a cooling plate; however, for the demonstrators considered in Table 8.3, the *maximum* difference is less than 1°C, which presumably has a minimal effect on aging under normal use conditions. Far greater differences were measured in the theoretical pump power and the average module temperature: here, the differences between demonstrators are significant enough to influence aging and vehicle range (also through the energy consumption of the fluid pump).

The experimental results demonstrate that some sort of BTMS is required for a high-performance PHEV in order to guarantee performance and system lifetime at extreme use cases. Due to the high heat loss, the heat capacity of the cells alone is insufficient for controlling the maximum temperature, and correspondingly, the high heat capacity determines the reaction time of the selected BTMS. Even if the vehicle is not operated under adverse conditions, a BTMS can pre-condition the battery cells and even maintain the ideal storage temperature when the vehicle is not in use. The experimental data shows that thermal performance is linked to the contact area of the coolant with the cell: a greater contact area increases heat transfer and simultaneously allows for a lower coolant flow rate. The current state of the art, a cooling plate, improves the average temperature of the cells versus no thermal management system, especially in adverse use cases. The potential of heat pipes has been demonstrated in this research: even in a (thermally) sub-optimal layout, the performance of the heat pipe technology demonstrator approaches that of the state of the art (cooling plate) while transferring the heat a much greater distance.

The pressure drops over all demonstrators have optimization potential. For all applications involving liquid coolant, the inlet and outlet geometries must be considered, as the pressure loss is often more significant than over the BTMS itself. Heat pipes themselves require no external energy input, resulting in a high potential for future energy-saving applications when combined with an optimized heat exchanger.

The potential vehicle suitability of all concepts is very similar in terms of weight and volume; the differences lie in the dimension that is increased versus a module without thermal management. Thus, in terms of vehicle suitability, the ideal BTMS depends on the location of the battery system: if height is critical, then a spacer-setup is desirable, but if length is more critical, a cooling plate is ideal.

The producibility and economic viability is in a most general sense a function of the production volume. To maximize production volume, all degrees of vehicle electrification (mild-hybrid, plug-in hybrid, full electric, etc.) should be combined. The ideal cell type for each application will be different, but based on the trend to standardized dimensions (Section 1), the external geometry and dimensions of a module could remain constant.

For a module having fixed dimensions, a BTMS attached to the outside the module is preferable. Thus, different cells types could be included within the module, while the contact surface to the BTMS remained the same. Then, depending on the use case and/or cell type, the externally attached BTMS could be scaled accordingly (or even omitted). The common parts for the resulting "universal" module would then become more economical. From this viewpoint, spacers are then only practical for very high performance applications. Such projects (e.g. sports cars) typically have lower sales volumes and a higher price point, which make the implementation of a spacer feasible.

In terms of material cost, aluminum is favorable; however, the cost of electrically insulating films is currently quite high because of the small amounts utilized for traditional consumer electronics applications. Thus, if the module itself is not electrically insulated, ceramic profiles in the form of a cooling plate may provide a cost effective solution. In terms of system cost reduction, heat pipes show great potential: already implemented in various consumer electronics and vehicle applications, the mass-market viability has been proven. Further research into alternative material combinations for all BTMS types (i.e. metal and plastic) may provide greater opportunities for reduced cost and weight, especially for cooling plates. As shown in Section 2.4, the complete life cycle of all materials must be considered to maximize the global environmental benefit of vehicle electrification.

Based on this analysis, the recommendations for a high-performance PHEV are presented in the following section. The recommendations are based on optimal solutions identified in Sections 5 through 7, and consider the role of the BTMS as a structural component.

8.4. Design Recommendations for a, High-Performance PHEV

This work has shown the potential of various concepts and provided an insight into critical design features. Ultimately, a universal recommendation of the "best" BTMS is not feasible due to the various vehicle applications possible. However, from the viewpoint of this research, using a standardized, pre-assembled module (mechanical compression, electrical insulation between cells, busbars, etc.) across all applications—and then integrating an appropriately scaled BTMS—is desirable. To facilitate this reduction in production complexity, the BTMS must then contact outside the module, and not within. Additionally, safety should be enhanced by separating the cells from the coolant when possible. Considering the potentials discussed in this research, two BTMS designs are ultimately recommended:

- For structural integration, a cast battery system casing with coolant channels sealed from the outside (described in **Section 5.3, Figure 5.10**), and
- When structural integration is not desired, a cooling plate consisting of a spreader plate and heat pipes (described in **Section 7.3, Figure 7.8**).

These layouts can be scaled to meet the required heat transfer depending on the cell type and thus the predicted cell heat loss. In this way, a single, scalable BTMS can be combined with an adaptable, universal module geometry for all vehicle electrification projects.

9. Conclusion

The need for affordable electric vehicles with will become more pressing as new emissions regulations take effect. Increasing the lifetime and reducing the cost of the battery system is crucial. To help achieve this goal, optimal cell operating temperatures must be facilitated so that vehicle range and system lifetime increase while reducing battery system size and cost. As a result, novel thermal management concepts must be identified and analyzed to determine the ideal solution in terms of thermal performance, energy consumption, vehicle suitability, producibility and economic viability. This research presents a method for efficiently and reproducibly analyzing diverse battery thermal management concepts in an early stage of development to assist in battery system design. The high spatial resolution facilitates analyses not previously available in the literature. The gathered data allows for the derivation of mathematical models for various thermal management systems based on the first law of thermodynamics, and provides design guidelines for battery thermal management system design. The methodology and analysis were demonstrated for a PHEV-2 cell, but can also be applied to other cell types.

The major points presented are summarized below:

- The SBC is a novel method to analyze battery thermal management systems (BTMS). By modeling the cell without the use of active chemistry, reproducible experimental trials are possible in an early stage of BTMS development with virtually no safety risks. The rapid reconditioning made possible by the internal cooling circuit allows for more trials to be performed and for hardware-in-the-loop integration. The sixteen thermocouples per cell (128 in the module) integrated within the cell casing provide an undisturbed thermal contact surface and a special resolution not currently found in the literature. The eight heating pads per cell allow for the simulation of extreme conditions or the simulation of damaged or aged cells.

- The developed and adapted analysis criteria provide a multifactorial comparison between new thermal management concepts and a module without thermal management. Many indications of technical performance can thereby be quantified and more precisely compared.

- The analysis of production techniques for cooling plates with liquid coolant show that welding techniques such as friction-stir welding (FSW), electron beam welding (EBW), BondWELD, and adhesion allow for weight-saving material combinations (i.e. different alloys or types), structural integration, and a physical separation between the cells and the coolant to increase crash safety. Analyses on channel geometry demonstrate the significance of convective heat transfer (channel footprint) to the coolant versus conduction in the solid medium. The induced temperature gradients are less critical in relation to the improvement of the average module temperature through a cooling plate. Pressure drop is far more sensitive to geometrical changes—especially at the inlet and outlet—than is the thermal behavior of the cells.

- Actively-streamed spacers between cells provide the ultimate heat removal, further demonstrating the dominance of heat transfer to coolant versus conduction in the solid medium at the scale analyzed. The weight and cost savings of positioning a spacer every second cell (versus between each cell) facilitate similarly high heat transfer. Due to the large contact surface, the coolant flow rate can be lowered versus a cooling plate. Alternative materials, such as ceramic, provide good thermal performance with the added benefit of increased compressive strength, durability, corrosion resistance and electrical insulation. Due there properties, ceramic should also be considered for cooling plate applications.

- Heat pipes display a high potential for BTMS applications. In terms of vehicle integration, the use of heat pipes to remove heat from the module (spacer or cooling plate) is currently the most advantageous layout. The opposing side can then be connected to an optimal heat exchanger. The current large-scale implementation of heat pipes in the consumer electronics industry indicates good producibility and economic viability. The number and size of the heat pipes can be scaled within a BTMS in order to meet the thermal demands of various cell / module types. Alternative heat pipe layouts (such as flat heat pipes) as well as more complex designs using additive production techniques will further enhance the attractiveness of heat pipes.

- The inter-concept comparison of diverse thermal management concepts shows that active spacers should be used only when high heat loads are anticipated, because many conventional applications do not require the high thermal conductance offered by a spacer solution. As a result, a universal module that can accommodate different cell types (HEV, PHEV, BEV, etc.) should be combined with a cooling plate that can also be scaled to meet the thermal requirements of the application. As a result, the producibility and economic viability can be maximized through the use of common parts, while battery system lifetime and performance are enhanced.

The findings of this research provide a framework for BTMS design. The fitted solutions to the first law of thermodynamics can be implemented into other thermal models. The presented method can be utilized and adapted to analyze new systems. The full potential of the SBC has however not been shown within the scope of this research.

Suggestions for Improvement and Outlook

The integration of a dynamic cell model with instantaneous temperature feedback and the full Hardware-in-the-Loop (HiL) functionality presents a further application of the presented method. A dynamic control system for the coolant properties provides dynamic feedback for total vehicle thermal management simulations, and assists in dynamically optimizing the control system in order to minimize energy consumption and maximize vehicle range.

The composite heat loss profile used was a compromise to maximize experimental efficiency; however, comparing the thermal management concepts at additional drive cycles would solidify the comparison in terms of the robustness of the various systems. Utilizing the multiple heat zones foreseen in the SBC would additionally allow for an analysis of how a BTMS responds to aged or damaged cells.

This research did not utilize structural simulations or experiments to optimize cooling plates. A beneficial extension of this research would be a co-simulation involving thermal behavior, pressure drop, and structural design for developing and optimizing structurally integrated cooling plates. In this way, design for life cycle optimization could also be more thoroughly integrated. The step of testing prototypes experimentally, as performed in this research, should however not be neglected.

An additional extension of this research would be the inclusion of the simulation of deformation during the production process. Especially when considering forms or welding, the cooling plate layout could be optimized to ensure that no additional production steps are required to achieve the required surface flatness.

Because an electric vehicle is meant to reduce the carbon footprint of the automotive industry, life-cycle engineering from the sourcing of the materials, through vehicle production and operational life must be considered. Potential second-life uses of vehicle batteries are a part of this solution.

Ultimately, the presented methodology and analysis have highlighted critical design features and provided a roadmap for the future multifactorial optimization of battery thermal management systems.

A. References

Ameli 2012

M. Ameli, B. Agnew, P. Leung, B. Ng, C. Sutcliffe, J. Singh, and R. McGlen, "A novel method for manufacturing sintered aluminum heat pipes (SAHP)," Applied Thermal Engineering, vol. 52, no. 2, pp. 498–504. DOI: 10.1016/j.applthermaleng.2012.12.011

Annaratone 2010

D. Annaratone, *Engineering Heat Transfer*. Springer, 2010. DOI 10.1007/978-3-642-03932-4_8

Arena 2014

F. Arena, L. Mezzana, A. Doyon, H. Suzuki, K. Lee, T. Becker, The Automotive CO2 Emissions Challenge: 2020 Regulatory Scenario for Passenger Cars. 2014. Available: http://www.adlittle.com/downloads/tx_adlreports/ADL_AMG_2014_Automotive_CO2_Emissions_Challenge.pdf

Arab 2014

M. Arab and A. Abbas, "A model-based approach for analysis of working fluids in heat pipes," Applied Thermal Engineering, vol. 73, no. 1, pp. 751–763. DOI: 10.1016/j.applthermaleng.2014.08.001

Audi 2014

AUDI AG, 2014

Bandhauer 2011

T. M. Bandhauer, S. Garimella, and T. F. Fuller, "A Critical Review of Thermal Issues in Lithium-Ion Batteries," J. Electrochem. Soc, vol. 158, no. 3, pp. R1-R25. DOI: 10.1149/1.3515880

Belyaev 2014

A. Belyaev, D. Fedorchenko M. Khazhmuradov, A. Lukhanin, O. Lukhanin, Y. Rudychev, B. Khalighi, T. Han, E. Yen, and U. S. Rohatgi, "Investigation of Heat Pipe Cooling of L-Ion Batteries." Brooklin National Laboratory BNL-105291-2014-CP. Available: https://www.bnl.gov/isd/documents/86095.pdf

Bernhard 2014

F. Bernhard, *Handbuch der Technischen Temperaturmessung*. Springer, 2014

Bhomik 2015

S. Bhowmik, R. Benedictus, and Y. Dan, "Adhesive Bonding Technology," in *Handbook of Manufacturing Engineering and Technology*, A. Y. C. Nee, Springer, 2015, pp. 765–784

BMBF 2014

Research Project eproduction. BMBF (German Ministry of Education and Research) project ID: 13N12027. 2011 to 2014. Contributors citied in this work: Andreas Roth, iwb of the TU-Munich; Marc Essers, ISF of the RWTH Aachen; Andreas Track and Jörg Demaske, FEES Verzahnungstechnik; Christian Schuch and Uwe Maurischat, Fraunhofer IFAM Bremen

Boddakayala 2013

B. Boddakayala, "Cooling System for Vehicles Batteries." U.S. Patent US201130183555A1. Issued July 18, 2013

Buford 2011

K. Buford, J. Williams, and M. Simonini, "Determining Most Energy Effcient Cooling Strategy of a Rechargable Energy Storage System," SAE International, vol. 2011-01-0893. DOI: 10.4271/2011-01-0893

REFERENCES

c2es 2014

Center for Climate and Energy Solutions, Federal Vehicle Standards.
Available: http://www.c2es.org/federal/executive/vehicle-standards

Carduso 2013

J. Carduso, "Kühlsystem mit aktiver Kabinenlüftung für eine Fahrzeugbatterie."
German Patent DE102013210937A1. Issued June 12, 2013

Cengal 2007

Y. Cengel. Heat and Mass Transfer: A Practical Approach. 3rd ed. N.p.:
McGraw-Hill Companies, 2007

Ceramtec 2015

Ceramtec GmbH, CeramCool: the Ceramic System for High Power Packaging.
Available:
https://www.ceramtec.de/files/el_ceramcool_the_ceramic_heat_sink_en.pdf
Last referenced January 2015

Chen 2012

Y.T. Chen, S.W. Kang, Y.H. Hung, C.H. Huang, and K.C. Chien, "Feasibility
study of an aluminum vapor chamber with radial grooved and sintered powders
wick structures," Applied Thermal Engineering, vol. 51, no. 1-2, pp. 864–870.
DOI: 10.1016/j.applthermaleng.2012.10.035

Childs 2003

P. Childs, "Advances in temperature measurement," Advances in Heat Transfer,
vol. 36, pp. 111–181. DOI: 10.1016/S0065-2717(02)80006-5

Connors 2009

M. J. Connors and J. A. Zunner, "The Use of vapor chambers and heat pipes for
cooling military embedded electronic devices," in MILCOM 2009 IEEE Military
Communications Conference. Boston, MA, USA

Daubitzer 2012

N. Daubitzer, M. Engelhart, T. Heckenberger, and T. Himmer, "Device for
Pressing a cooler against a Battery." Patent WO2012/104394A1. Issued August
9, 2012

DIN 8580

Fertigungsverfahren - Begriffe, Einteilung, DIN 8580.

Dunn 1982

P. D. Dunn and D. A. Reay, Heat pipes, 3rd ed. Oxford, New York: Pergamon
Press, 1982

Ebnesajjad 2011

S. Ebnesajjad, Handbook of adhesives and surface preparation. Elsevier, op.
2011

Eden 2012

K. Eden and H. Gebhard, Eds, Dokumentation in der Mess- und Prüftechnik:
Messen - Auswerten - Darstellen Protokolle - Berichte - Präsentationen.
Wiesbaden: Vieweg+Teubner Verlag / Springer Fachmedien Wiesbaden GmbH,
Wiesbaden, 2012

Eireiner 2006

D. R. Eireiner, Prozessmodelle zur statischen Auslegung von Anlagen für das
Friction Stir Welding. München: Utz, 2006

El-Mahallawi 2013

I. El-Mahallawi and S. El-Raghy. Recycling of Metal Products. Ed. A.
Richardson, Reuse of Materials and Byproducts in Construction, Green Energy
and Technology, pp Springer 2013, DOI: 10.1007/978-1-4471-5376-4_3

EPA 1999

United States Environmental Protection Agency, The History of Reducing Tailpipe Emissions. EPA420-F-99-017. May 1999. Available: http://www.epa.gov/oms/consumer/f99017.pdf

EPA 2009

United States Environmental Protection Agency, Clean Air Act Mobile Source Civil Penalty – Vehicle and Engine Certification Requirements. January 2009. Available: https://www.epa.gov/sites/production/files/documents/vehicleengine-penalty-policy_0.pdf

Fleckenstein 2012

M. Fleckenstein, O. Bohlen, and B. Bäker, "Aging Effect of Temperature Gradients in Li-ion Cells Experimental and Simulative Investigations and the Consequences on Thermal Battery Management," The 26th International Battery, Hybrid and Fuel Cell Electric Vehicle Symposium. Los Angeles, California, USA, May. 6, 2012

Fletcher 1988

L. S. Fletcher, "Recent Developments in Contact Conductance Heat Transfer," J. Heat Transfer, vol. 110, no. 4b, p. 1059. DOI: 10.1115/1.3250610

Gawthrop 2013

P. Gawthrop, "Climate Control System for Hybrid Vehicles using Thermoelectric Devices." US Patent 2013/0327063A1. Issued December 12, 2013

Ghassemali 2015

E. Ghassemali, X. Song, M. Zarinejad, A. Danno, and M. J. Tan, "Bulk Metal Forming Processes in Manufacturing," in *Handbook of Manufacturing Engineering and Technology*, A. Y. C. Nee, Springer, 2015, pp. 171–230

Giuliano 2012

M. R. Giuliano, A. K. Prasad, and S. G. Advani, "Experimental study of an air-cooled thermal management system for high capacity lithium–titanate batteries," Journal of Power Sources, vol. 216, pp. 345–352. DOI: 10.1016/j.jpowsour.2012.05.074

Groll 2002

M. Groll and S. Khandekar, "Pulsating Heat Pipes: A Challenge and Still Unsolved Problem in Heat Pipe Science," in HEAT 2002: The Third International Conference on Transport Phenomena in Multiphase Systems, Baranów Sandomierski, Poland, June 24-27, 2002, pp. 35–43

Grote 2009

K.H. Grote and E.K. Antonsson, *Springer Handbook of Mechanical Engineering*. Springer, 2009

Guenther 2012

H.O. Günther and H. Tempelmeier, *Produktion und Logistik*. Springer, 2012

Guo 2015

J. Guo, "Solid State Welding Processes in Manufacturing Welding process Manufacturing, solid state welding processes" in *Handbook of Manufacturing Engineering and Technology*, A. Y. C. Nee, Springer, 2015, pp. 569–592

Hallaj 2000

S. Al Hallaj, J. Prakash, and J. Selman, "Characterization of commercial Li-ion batteries using electrochemical–calorimetric measurements," Journal of Power Sources, vol. 87, no. 1-2, pp. 186–194. DOI: 10.1016/S0378-7753(99)00472-3

REFERENCES

Held 2011

M. Held and M. Baumann, "Assessment of the Environmental Impacts of Electric Vehicle Concepts," in Towards Life Cycle Sustainability Management, M. Finkbeiner (ed.), pp. 535-546, Springer 2011. DOI: 10.1007/978-94-007-1899-9_52

Heubner 2013

C. Heubner, M. Schneider, C. Lämmel, U. Langklotz, and A. Michaelis, "In-operando temperature measurement across the interfaces of a lithium-ion battery cell," Electrochimica Acta, vol. 113, pp. 730–734. DOI: 10.1016/j.electacta.2013.08.091

Hirsch 2013

S. Hirsch, C. Schmid, A. Wiebelt, and M. Stripf, "Thermisch Übergangsvorrichtung, Temperierplatte und Energiespeichereinheit." German Patent DE102011084002A1. Issued April 4, 2013

Hofmann 2014

P. Hofmann, "Hybridkomponenten," in *Hybridfahrzeuge*, P. Hofmann, Springer, 2014, pp. 145–286

Hornbogen 2008

E. Hornbogen, G. Eggeler, and E. Werner, "Keramische Werkstoffe," in *Springer-Lehrbuch*, Werkstoffe, E. Hornbogen, G. Eggeler, and E. Werner, Springer, 2008, pp. 315–344

Huelsenberg 2014

D. Huelsenberg, *Keramik*. Springer, 2014. DOI: 10.1007/978-3-642-53883-4

Huehner 2012

S. Huehner, A. Dillmann, and R. Holp, "Storage Unit for Storing Electrical Energy with a Heat Pipe." Patent WO 2012/175300A1. Issued December 27, 2012

ICCT 2014

International Council on Clean Transportation, EU CO2 Emission Standards for Passenger Cars and Light-Commercial Vehicles. 2014. Available: http://www.theicct.org/sites/default/files/publications/ICCTupdate_EU-95gram_jan2014.pdf

IEA 2007

Tracking Industrial Energy Efficiency and CO2 Emissions (2007). International Energy Agency (IEA), Head of Communication and Information Office, France.

IEC 2007

Thermocouples - Part 3: Extension and compensating cables - Tolerances and identification system, International Electrotechnical Commission. IEC 60584-3:2007

Inuoue 2014

Y. Inoue, "Battery Unit with Blower." US Patent 8906530B2. Issued December 9, 2014

Isermeyer 2010

T. Isermeyer, H. Damsohn, and C. Pfender, "Elektrochemische Energiespeichereinheit mit Kühlvorrichtung. " European Patent EP2153487B1. Issued February 17, 2010

Jayaraman 2011

S. Jayaraman, G. Anderson, S. Kaushik, and K. Philip, "Modeling of Battery Pack Thermal Management for a Plug-In Hybrid Electric Vehicle," SAE International, vol. 2011-01-0666. DOI: 10.4271/2011-01-0666

Kampker 2014

A. Kampker, "Die Produktion des Hochvolt-Speichersystems," in *Elektromobilproduktion*, A. Kampker, Springer, 2014, pp. 43–113

Khandkar 2003

M. Z. H. Khandkar, J. A. Khan, and A. P. Reynolds, "Prediction of temperature distribution and thermal history during friction stir welding: input torque based model," Science and Technology of Welding and Joining, vol. 8, no. 3, pp. 165–174. DOI: 10.1179/136217103225010943

Kim 2007

U. S. Kim, C. B. Shin, and C.-S. Kim, "Effect of electrode configuration on the thermal behavior of a lithium-polymer battery," Journal of Power Sources, vol. 180, no. 2, pp. 909–916. DOI: 10.1016/j.jpowsour.2007.09.054.

Koehler 2013

U. Koehler, "Aufbau von Lithium-Ionen-Batteriesystemen," in *Handbuch Lithium-Ionen-Batterien*, R. Korthauer, Springer, 2013, pp. 95–106

KPMG 2010

KPMG International Cooperative, The Transformation of the Automotive Industry: The Environmental Regulation Effect. 2010. Available: https://www.kpmg.com/US/en/industry/Japanese-Practice/Documents/2010-issue2/transformation-automotive-industry.pdf

Krinke 2011

S. Krinke, "Implementing Life Cycle Engineering Efficiently into Automotive Industry Processes," in Towards Life Cycle Sustainability Management, M. Finkbeiner (ed.), pp. 535-546, Springer 2011. DOI: 10.1007/978-94-007-1899-9_55

Kritzer 2013

P. Kritzer and O. Nahrwold, "Dichtungs- und Elastomerkomponenten für Lithium-Batteriesysteme," in *Handbuch Lithium-Ionen-Batterien*, R. Korthauer, Springer, 2013, pp. 119–129

Kwak 2013

J. W. Kwak, K. H. Song, H. S. Lee, and B. S. Choi, "Radient Heat Plate for Battery Cell Module and Battery Cell Module having the Same." U.S. Patent US20130202924A1. Issued August 8, 2013

Lamp 2013

P. Lamp, "Anforderungen an Batterien für die Elektromobilität," in *Handbuch Lithium-Ionen-Batterien*, Ed. R. Korthauer, Springer, 2013, pp. 393–415

Legrand 2013

N. Legrand, B. Knosp, P. Desprez, F. Lapicque, and S. Raël, "Physical characterization of the charging process of a Li-ion battery and prediction of Li plating by electrochemical modelling," Journal of Power Sources, vol. 245, pp. 208–216. DOI: 10.1016/j.jpowsour.2013.06.130

Li 2013

Z. Li, J. Zhang, B. Wu, J. Huang, Z. Nie, Y. Sun, F. An, and N. Wu, "Examining temporal and spatial variations of internal temperature in large-format laminated battery with embedded thermocouples," Journal of Power Sources, vol. 241, pp. 536–553. DOI: 10.1016/j.jpowsour.2013.04.117

Liaw 2003

B.Y. Liaw, E.P. Roth, R.G. Jungst, G. Nagasubramanian, H.L. Case, and D.H. Doughty, "Correlation of Arrhenius behaviors in power and capacity fades with cell impedance and heat generation in cylindrical lithium-ion cells," Journal of Power Sources, vol. 119-121, pp. 874–886. DOI: 10.1016/S0378-7753(03)00196-4

Lohwasser 2009

D. Lohwasser and Z. Chen, Friction stir welding: From basics to applications. Cambridge: Woodhead Publishing, 2009

REFERENCES

Lukhanin 2012

A. Lukhanin, A. Belyaev, D. Fedorchenko, M. Khazhmuradov, O. Lukhanin, Y. Rudychev, and U. S. Rohatgi, "Thermal Characteristics of Air Flow Cooling in the Lithium Ion Batteries Experimental Chamber," in Proceedings of the ASME summer heat transfer conference 2012: HT 2012, pp. 129–133

Lukhanin 2013

A. Lukhanin, A. Byelyayev, D. Fedorchenko, M. Khazhmuradov, O. Lukhanin, S. Martynov, Y. Rudychev, E. Sporov, and U. S. Rohatgi, "Investigation of Air Flow Cooling of Li-Ion Batteries," in Proceedings of the ASME International Mechanical Engineering Congress and Exposition 2013, New York, N.Y: American Society of Mechanical Engineers. DOI: 10.1115/IMECE2013-63095

Mahefkey 1971

E. T. Mahefkey and M. M. Kreitman, "An Intercell Planar Heat Pipe for the Removal of Heat During the Cycling of a High Rate Nickel Cadmium Battery," J. Electrochem. Soc, vol. 118, no. 8, p. 1382. DOI: 10.1149/1.2408330

Marretta 2012

L. Marretta, R. Di Lorenzo, F. Micari, J. Arinez, D. Dornfeld. Material Substitution for Automotive Applications: A Comparative Life Cycle Analysis. In Proceedings of the 19th CIRP Conference on Life Cycle Engineering, University of California at Berkeley, USA, May 23-25, 2012., pp 61-66, Springer 2012. DOI: 10.1007/978-3-642-29069-5_11

Meisel 2014

P. Meisel, M. Jobst, W. Lippmann, and A. Hurtado, "Design and manufacture of ceramic heat pipes for high temperature applications," Applied Thermal Engineering, vol. 75, pp. 692–699. DOI: 10.1016/j.applthermaleng.2014.10.051

Michaelis 2009

A. Michaelis, "Ceramics," in *Technology Guide*, Ed. H.-J. Bullinger, Springer 2009, pp. 14-17. ISBN: 978-3-540-88546-7

Mishra 2007

R. S. Mishra and M. W. Mahoney, Friction stir welding and processing. Materials Park, Ohio: ASM International, 2007

Mutyala 2014

M.S.K. Mutyala, J. Zhao, J. Li, H. Pan, C. Yuan, and X. Li, "In situ temperature measurement in Lithium Ion Battery by transferable flexible thin film thermocouples // In-situ temperature measurement in lithium ion battery by transferable flexible thin film thermocouples," Journal of Power Sources, vol. 260, pp. 43–49. DOI: 10.1016/j.jpowsour.2014.03.004

Nagano 2001

Y. Nagano, S. Jogan, and T. Hashimoto, "Mechanical Properties of Aluminum Die Casting Joined by FSW," in Third International Symposium on Friction Stir Welding. Kobe, Japan, 2001

Navidi 2006

W. C. Navidi, Statistics for engineers and scientists. Boston, Mass: McGraw-Hill, 2006

Nelson 2000

T. Nelson, C. Sorensen, C. Johns, S. Strand, and J. Christensen, "Joining of Thermoplastics with Friction Steer Welding," in Second International Symposium on Friction Stir Welding. Goteborg, Sweden, 2000

Neumeister 2010

D. Neumeister, A. Wiebelt, and T. Heckenberger, "Systemeinbindung einer Lithium-Ionen-Batterie in Hybrid- und Elektroautos," ATZ Automobiltech Z, vol. 112, pp. 250–255.

NHTSA 2011
National Highway Traffic Safety Administration, Corporate Average Fuel Ecenomy (CAFE): Fact Sheet. December 2011. Available: www.nhtsa.gov/staticfiles/rulemaking/pdf/cafe/CAFE_2017-25_Fact_Sheet.pdf

Noie 2014
S. Noie, "Heat transfer characteristics of a two-phase closed thermosyphon," Applied Thermal Engineering, vol. 25, no. 4, pp. 495–506. DOI: 10.1016/j.applthermaleng.2004.06.019

Okawa 2013
H. Okawa, H. Kishita, T. Yamanaka, and M. Takeuchi, "Cell-Temperature Adjustment Device." Patent WO2013114513A1. Issued December 26, 2013

Park 2009
Y. Park, S. Jun, S. Kim, and D.-H. Lee, "Design optimization of a loop heat pipe to cool a lithium ion battery onboard a military aircraft," J Mech Sci Technol, vol. 24, no. 2, pp. 609–618. DOI: 10.1007/s12206-009-1214-6

Pesaran 2003
A. Pesaran, T. Stuart, and A. Vlahinos, "Cooling and Preheating of Batteries in Hybrid Electric Vehicles," The 6th ASME-JSME Thermal Engineering Joint Conference, vol. TED-AJ03-633

Pesaran 2006
A. Pesaran and G.-H. Kim, "Battery Thermal Management System Design Modeling,"22nd International Battery, Hybrid and Fuel Cell Electric Vehicle Conference and Exhibition (EVS-22). Yokohama, Japan, Oct. 23, 2006

Pesaran 2013
A. Pesaran, M. Keyser, K. Smith, G.H. Kim, and S. Santhanagopalan, "Tools for Designing Thermal Management of Batteries in Electric Drive Vehicles," Large Lithium Ion Battery Technology & Application Symposia Advanced Automotive Battery Conference. Pasadena, CA, USA, Feb. 4, 2013

Pettinger 2013
K.-H. Pettinger, "Prüfverfahren in der Fertigung," in *Handbuch Lithium-Ionen-Batterien*, R. Korthauer, Springer, 2013, pp. 259–267

Qu 2006
W. Qu and H. Ma, "Theoretical analysis of startup of a pulsating heat pipe," International Journal of Heat and Mass Transfer, vol. 50, no. 11-12, pp. 2309–2316. DOI: 10.1016/j.ijheatmasstransfer.2006.10.043

Raijmakers 2013
L. Raijmakers, D. Danilov, J. van Lammeren, M. Lammers, and P. Notten, "Sensorless battery temperature measurements based on electrochemical impedance spectroscopy," Journal of Power Sources, vol. 247, pp. 539–544. DOI: 10.1016/j.jpowsour.2013.09.005

Rao 2011
Z. Rao and S. Wang, "A review of power battery thermal energy management," Renewable and Sustainable Energy Reviews, vol. 15, no. 9, pp. 4554–4571. DOI: 10.1016/j.rser.2011.07.096

Rao 2012
Z. Rao, S. Wang, M. Wu, Z. Lin, and F. Li, "Experimental investigation on thermal management of electric vehicle battery with heat pipe," Energy Conversion and Management, vol. 65, pp. 92–97. DOI: 10.1016/j.enconman.2012.08.014

Reay 2014a
D. Reay, P. Kew, and R. McGlen, "Heat pipe components and materials," in Heat Pipes: Elsevier, 2014, pp. 65–94. DOI: 10.1016/B978-0-08-098266-3.00003-0

Reay 2014b

D. Reay, P. Kew, and R. McGlen, "Heat pipe manufacture and testing," in Heat Pipes: Elsevier, 2014, pp. 105–133. DOI:10.1016/B978-0-08-098266-3.00005-4

Reisgen 2012

U. Reisgen, M. Schleser, and M. Essers, "BondWELD – Klebstofffixiertes Rührreibschweißen von dünnen Aluminiumblechen," in DVS-Berichte, vol. 295, 33. Assistentenseminar Füge- und Schweißtechnik: Vorträge der gleichnamigen Veranstaltung in Goslar vom 6. bis 8. September 2012, DVS Media GmbH, Ed, 2013, pp. 120–125

Remmen 2007

A. Remmen, A.- A. Jensen, J. Frydendal, Life Cycle Management: A Business Guide to Sustainability (2007). United Nations Environment Program (UNEP) and Danish Standards, pp. 12. ISBN: 978-92-807-2772-2

Robinson 2013

J. B. Robinson, J. A. Darr, D. S. Eastwood, G. Hinds, P. D. Lee, P. R. Shearing, O. O. Taiwo, and D. J. Brett, "Non-uniform temperature distribution in Li-ion batteries during discharge – A combined thermal imaging, X-ray micro-tomography and electrochemical impedance approach," Journal of Power Sources, vol. 252, pp. 51–57. DOI: 10.1016/j.jpowsour.2013.11.059

Ruys 2015

A. Ruys, O. Gingu, G. Sima, and S. Maleksaeedi, "Powder Processing of Bulk Components in Manufacturing," in *Handbook of Manufacturing Engineering and Technology*, A. Y. C. Nee, Springer, 2015, pp. 487–566

Schmidt 2013

J. P. Schmidt, S. Arnold, A. Loges, D. Werner, T. Wetzel, and E. Ivers-Tiffée, "Measurement of the internal cell temperature via impedance: Evaluation and application of a new method," Journal of Power Sources, vol. 243, pp. 110–117. DOI: 10.1016/j.jpowsour.2013.06.013

Schneider 2013

C. Schneider. (2013). Analysen und Untersuchungen von Kühl- und Dichtsystemen bei Hochvolt-Batteriesystemen unter Berücksichtigung der Produktionsanforderungen für eine Serienfertigung. Unpublished master's thesis, Wilhelm Büchner Hochschule (Fachbereich Mechatronik), Darmstadt,Germany

Schubert 2010

G. Schubert, Electron Beam Welding: Process, Applications and Equipment. Available: http://www.ptreb.com/Customer-Content/WWW/CMS/files/Electron_Beam_Welding_Process_Applications_and_Equipment2.pdf

Schultz 1993

H. Schultz, *Electron beam welding*. Cambridge, England: Abington, 1993.

Singh 2007

R. Singh, A. Akbarzadeh, and M. Mochizuki, "Operational characteristics of a miniature loop heat pipe with flat evaporator," International Journal of Thermal Sciences, vol. 47, no. 11, pp. 1504–1515. DOI: 10.1016/j.ijthermalsci.2007.12.013

Smith 2014

J. Smith, M. Hinterberger, P. Hable, and J. Koehler, "Simulative method for determining the optimal operating conditions for a cooling plate for lithium-ion battery cell modules," Journal of Power Sources, vol. 267, pp. 784–792. DOI: 10.1016/j.jpowsour.2014.06.001

Smith 2015
J. Smith, M. Hinterberger, C. Allmann, C. Budde, S. Schüren, J. Köhler. "Experimental optimization of thermal management systems for high-voltage batteries using Simulation." Batterieforum Deutschland, Berlin, Germany, 21-23 January 2015

Song 2013
C. H. Song, "Battery Cooling System using Thermoelectric Element." Patent WO2013/137612A1. Issued September 19, 2013

Sterner 2014a
M. Sterner and I. Stadler, "Speicherbedarf im Verkehrssektor," in *Energiespeicher - Bedarf, Technologien, Integration*, M. Sterner and I. Stadler, Springer, 2014, pp. 141–159

Sterner 2014b
M. Sterner and I. Stadler, "Elektrochemische Energiespeicher," in *Energiespeicher - Bedarf, Technologien, Integration*, M. Sterner and I. Stadler, Springer, 2014, pp. 197–294

Stripf 2012
M. Stripf, M. Wehowski, C. Schmid, and A. Wiebelt, "Thermomanagement von Hochleistungs-Li-Ionen-Batterien," ATZ Automobiltech Z, vol. 114, no. 1, pp. 52–57. DOI: 10.1365/s35148-012-0248-8

Sun 1996
Z. Sun and R. Karppi, "The application of electron beam welding for the joining of dissimilar metals: an overview," Journal of Materials Processing Technology, vol. 59, no. 3, pp. 257–267. DOI: 10.1016/0924-0136(95)02150-7

Syed 2013
S. A. Syed, R. J. Heydal, and J. G. Dorrough, "Wave Fin Battery Module." U.S. Patent US20130101881A1. Issued April 25, 2013

Tang 1998
W. Tang, X. Guo, J. C. McClure, L. E. Murr, and A. Nunes, "Heat Input and Temperature Distribution in Friction Stir Welding," Journal of Materials Processing & Manufacturing Science, vol. 7, no. 2, pp. 163–172. DOI: 10.1106/55TF-PF2G-JBH2-1Q2B

Teng 2011
H. Teng, Y. Ma, K. Yeow, and M. Thelliez, "Thermal Characterization of a Li-ion Battery Module Cooled through Aluminum Heat-Sink Plates," SAE International Journal of Passenger Cars, pp. 1331–1342. DOI: 10.4271/2011-01-2248

Toepfer 2013
S. Toepfer, "Flüssigkeitswärmetauscher aus Kunstoff für ein Batteriekühlsystem." German Patent DE102013200859A1. Issued August 1, 2013

Tran 2013
T.-H. Tran, S. Harmand, B. Desmet, and S. Filangi, "Experimental investigation on the feasibility of heat pipe cooling for HEV/EV lithium-ion battery," Applied Thermal Engineering, vol. 63, no. 2, pp. 551–558. DOI: 10.1016/j.applthermaleng.2013.11.048

Vetter 2005
J. Vetter, P. Novák, M. Wagner, C. Veit, K.-C. Möller, J. Besenhard, M. Winter, M. Wohlfahrt-Mehrens, C. Vogler, and A. Hammouche, "Ageing mechanisms in lithium-ion batteries," Journal of Power Sources, vol. 147, no. 1-2, pp. 269–281. DOI: 10.1016/j.jpowsour.2005.01.006

Walser 2014
D. Walser and B. Fragniere, "Battery Pack for a Motor Vehicle." Patent W2014/198778A1. Issued December 18, 2014

Wan 2014

Z. Wan, J. Deng, B. Li, Y. Xu, X. Wang, and Y. Tang, "Thermal performance of a miniature loop heat pipe using water–copper nanofluid," Applied Thermal Engineering, vol. 78, pp. 712–719. DOI: 10.1016/j.applthermaleng.2014.11.010

Wang 2014

Q. Wang, B. Jiang, Q. Xue, H. Sun, B. Li, H. Zou, and Y. Yan, "Experimental investigation on EV battery cooling and heating by heat pipes," Applied Thermal Engineering. DOI: 10.1016/j.applthermaleng.2014.09.083

Weileder 2012

S. Weileder, S. Köster, R. Löffler, R. Lustig, and A. Meijering, "Device for Supplying Power, Having a Cooling Assembly." Patent WO2012013315A1. Issued, February 2, 2012

Wiebelt 2009

A. Wiebelt, T. Isermeyer, T. Siebrecht, and T. Heckenberger, "Thermomanagement von Lithium-Ionen-Batterien," ATZ Automobiltech Z, vol. 111, pp. 500–504

Wu 2002

M.S. Wu, K.H. Liu, Y.Y. Wang, and C.C. Wan, "Heat dissipation design for lithium-ion batteries," Journal of Power Sources, vol. 109, no. 1, pp. 160–166. DOI: 10.1016/S0378-7753(02)00048-4

Xuan 2004

Y. Xuan, Y. Hong, and Q. Li, "Investigation on transient behaviors of flat plate heat pipes," Experimental Thermal and Fluid Science, vol. 28, no. 2-3, pp. 249–255. DOI: 10.1016/S0894-1777(03)00047-5

Yang 2011

X. Yang, Y. Yan, and D. Mullen, "Recent developments of lightweight, high performance heat pipes," Applied Thermal Engineering, vol. 33-34, pp. 1–14. DOI: 10.1016/j.applthermaleng.2011.09.006

Yang 2013

G. Yang, C. Leitão, Y. Li, J. Pinto, and X. Jiang, "Real-time temperature measurement with fiber Bragg sensors in lithium batteries for safety usage," Measurement, vol. 46, no. 9, pp. 3166–3172. DOI: 10.1016/j.measurement.2013.05.027

Yellishettya 2011

M. Yellishettya, G. M. Mudda, P.G. Ranjitha, and A. Tharumarajah, Environmental life-cycle comparisons of steel production and recycling: sustainability issues, problems and prospects. J. of Environmental Science and Policy (14) pp 650-663. 2011. DOI: 10.1016/j.envsci.2011.04.008

Yu 2013

R. Yu, F. Dubief, and M. Origuchi, "Device for Cooling the Batteries of an especially Electric Vehicle and Vehicle Comprising such a Device." US Patent: 8584780B2. Issued November 19, 2013

Zeyen 2013

M. G. Zeyen and A. Wiebelt, "Thermisches Management der Batterie," in *Handbuch Lithium-Ionen-Batterien*, R. Korthauer, Springer, 2013, pp. 165–175

Zhao 2012

J. Zhao, H. Li, H. Choi, W. Cai, J. A. Abell, and X. Li, "Insertable thin film thermocouples for in situ transient temperature monitoring in ultrasonic metal welding of battery tabs," Journal of Manufacturing Processes, vol. 15, no. 1, pp. 136–140. DOI: 10.1016/j.jmapro.2012.10.002

B. Abbreviations and Variables

Abbreviations

BEV	Battery Electric (or Full Electric) Vehicle
BTMS	Battery Thermal Management System
CFD	Computational Fluid Dynamics
CID	Current Interrupt Devise
CNG	Compressed Natural Gas
CO_2	Carbon Dioxide
E5, E10, E85	Ethanol percentage in gasoline (5, 10, 85%)
EBW	Electron Beam Welding
EU	European Union
FSW	Friction Stir Welding
gCO_2	Gram Carbon Dioxide
GHG	Green House Gas
HEV	(Mild) Hybrid Electric Vehicle
HiL	Hardware in the Loop
LFP	Lithium Ion Phosphate
LMO	Lithium Manganese Oxide
LNG	Liquid Natural Gas
NEDC	New European Driving Cycle
NMC	Lithium Nickel Manganese Cobalt Oxide
OEM	Original Equipment Manufacturer
PHEV	Plug-In Hybrid Electric Vehicle
POM	Polyoxymethylene
R / RC-Circuit	Resistor / Resistor-Capacitor Circuit
SBC	Smart Battery Cell
SOC	State of charge
TM	Thermal Management
US	United States

Latin Variables

A	Area	$[m^2]$
Bi	Biot Number	[ ]
c (c_p, c_v)	Specific heat (at constant pressure, volume)	$[J^1\,kg^{-1}\,K^{-1}]$
D,d	Diameter	$[m]$
g	Acceleration due to gravity	$[m^1\,s^{-2}]$
Gr	Grashof Number	[ ]
$h_{convection}$	Coefficient of convective heat transfer	$[W^1\,m^{-2}\,K^{-1}]$
I	Current	$[A]$
i'''	Current per unit cell volume	$[A^1\,m^{-3}]$
k	Coefficient of conduction	$[W^1\,m^{-1}\,K^{-1}]$
L	Length	$[m]$
L_c	Characteristic Length	$[m]$
Nu	Nusselt Number	[ ]

P	Power	[W]
Δp	Pressure difference	[mbar]
Pr	Prandtl Number	[ ]
Q	Rate of heat loss / generation	[W^1]
q'''	Heat generation per unit volume	[W^1 m^{-3}]
R	Thermal resistance	[K^1 W^{-1}]
s_{cell}	Standard sample deviation of cell temperature	[K]
s_D, s_L, s_T	Diagonal, longitudinal, and transverse distance	[m]
T	Temperature	[K]
ΔT	Temperature difference	[K]
$\Delta(\Delta T)$	Difference in the temperature gradients of each cell	[K]
U_{OCV}, U_{Total}	Open circuit potential, Overall cell potential	[V]
$\dot{V}$	Volumetric flow rate	[L^1 min^{-1}]
V	Volume	[m^3]
W	Weight	[kg]

Greek Variables

β	Coefficient of thermal expansion	[K^{-1}]
μ	Dynamic viscosity	[Pa s]
ρ	Density	[kg^1 m^{-1}]
σ	Electrical conductivity	[S^1 m^{-1}]
ϕ	Local potential in the cell	[V]

C. Details of the Experimental Setup

Ambient boundary conditions must also be consistent and reproducible in order to eliminate additional variables and isolate the effect of the BTMS on the cells. Therefore, all experimental technology demonstrators are placed on a 25 mm Polyoxymethylene (POM) base, and all other surfaces are left exposed. The primary reason for this decision is to avoid errors caused by insulating experimental setups that are very distinct in their layout: the same thickness and coverage over the entire demonstrator cannot be guaranteed, resulting in varying heat fluxes. On the other hand, the boundary condition of natural convection chosen must be guaranteed by the climate chamber, and convection and edge effects on the technology demonstrator will be observed in the experiments. However, an exposed setup facilitates rapid conditioning to the initial conditions for the experimental trials. Furthermore, safety is increased by giving the operator visual access to the setup.

The primary technical challenge of the custom climate chamber is guaranteeing the natural convection boundary condition. This is achieved through the combination of a large heat exchanger, a large chamber volume, and a low air flow speed. Even at the peak cell heat loss tested, the 1.6 kW output of the module is removed by the 4 kW water-gas heat exchanger. A large air inlet (2 in Figure C.1) combined with the large climate chamber volume (45 times the experimental setup) results in a low required flow speed (0.04 to 0.08 m/s) to maintain the climate chamber temperature during normal cell operating conditions (heat loss of 10 to 20 W/cell).

To maintain the set temperature, the aforementioned heat exchanger, a fan, a condensation collector, and an electric heater are installed below the climate chamber in a separate enclosure (1 in Figure C.2). From here, the air flows through a manifold to the climate chamber inlet (2 in Figure C.2). The nearly square-meter climate chamber inlet is fitted with a perforated metal plate that distributes the flow across the opening. Air temperature is measured in the manifold, at the outlet from the chamber, and in the middle of the chamber. Connectors for the sensors used in the experimental trials are mounted flush in the wall in order to not influence the flow in the lower portion of the climate chamber where the test setup is placed. Additionally, the low flow speed and large chamber minimize thermal boundary layer effects from the walls, allowing for a homogeneous temperature distribution. In this way, reproducible boundary conditions can be very accurately created.

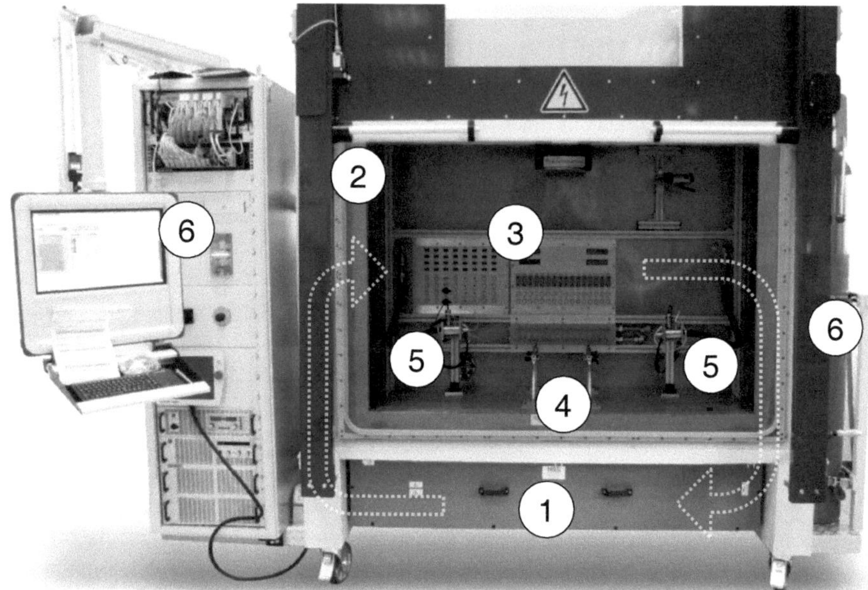

Figure C.1: Climate chamber overview: The climate chamber, showing the location of the climatization equipment (1), the 0.96 m² air-flow inlet (2), the connection panel for the sensors (3), BTMS mount (4), water-glycol circuit temperature and pressure measurement points (5). The white dashed arrows show the direction of low-speed air circulation.

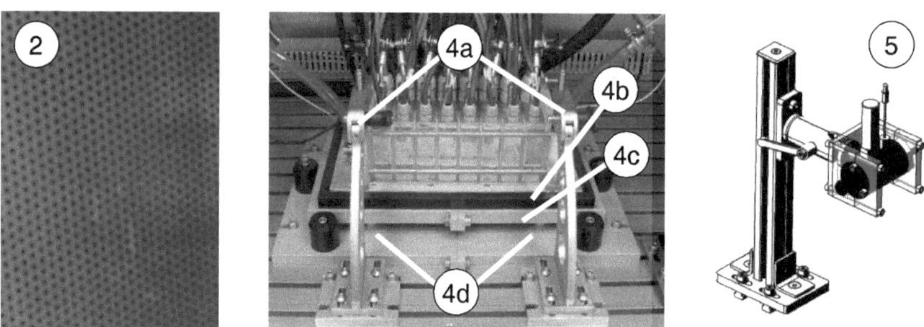

Figure C.2: Detail of select components: A detailed view of the perforated inlet (2), the BTMS mount for a cooling plate including the compression arms (4a), the 25mm POM insulating base (4b), the stiff-packing plate for the based (4c), and two of the force sensors (4d). The temperature and pressure measurement apparatus is shown in (5).

The selection of sensors and layout thereof is designed to meet both the goals of current experiments and potential extensions identified by the findings. Two PT100 temperature sensors as well as two pressure sensors are integrated into a coolant loop, and serve to measure coolant inlet temperature and pressure to a BTMS. Also on the coolant circuit is a flow rate regulator and magnetic flow meter to measure and set the flow rate. The coolant source and is a water-cooled process thermostat, allowing coolant temperatures from -20 to 60°C. Four force sensors are included to analyze the compression force between a module and cooling plate. The sensors are integrated into the BTMS mount, and can be utilized in combination with the compression arms to vary the compressive force. Hydraulic connectors for the SBC cooling circuit and 48 V power supply are located next to a connection panel for 32 Type-K sensors, as well as connectors for each of the 128 thermocouples. Because the climate chamber was designed to accommodate thermal management concepts of all shapes and sizes using both SBC and actual cells, the path from the thermocouple measurement point in the SBC to the data acquisition is foreseen with a connector. The thermocouple extension cables used in the climate chamber were obtained from the same strand (in cooperation with the distributer) as those integrated in the SBC, which serves to reduce measurement error. Finally, a complete calibration of the measurement system from SBC to the software was performed at five reference voltages corresponding to five temperatures, increasing accuracy across the desired temperature range.

D. Numerical Analysis of Cooling Plate Operating Conditions

The following numerical simulations were published in [Smith 2014], but the results regarding to the evaluation of the optimal operating conditions for a cooling plate are presented here.

The steady state solution of the temperature gradient of each individual cell $\Delta T_{cell,n}$, the difference of the individual cell temperature gradients of the module $\Delta(\Delta T_{cells})$, and the maximum module temperature T_{max} are the quantities of interest for battery system thermal behavior. The pressure drop along the cooling plate (Δp) determines the coolant circuit pumping power (BTMS energy consumption), pump size, and pump cost.

The interplay of the aforementioned quantities of interest are visually displayed by first making the quantities dimensionless--thereby allowing them to be simultaneously visualized--and second by plotting the dimensionless quantities versus coolant inlet temperature and volumetric flow rate. This allows a designer to optimize the vehicle thermal management system by selecting the ideal operating conditions and provides a dynamic input to the one-dimensional model. Furthermore, the data is plotted as a function of cell heat loss, reflecting various vehicle operating conditions.

The multi-objective analysis is demonstrated on a cooling plate consisting of a 0.5 mm top plate and 4 mm high fluid channel.

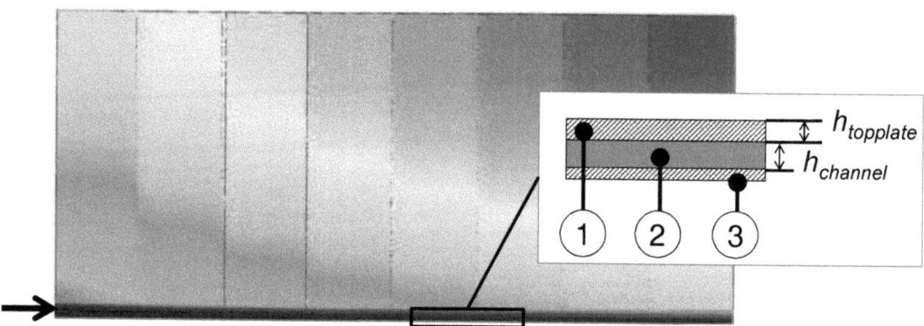

Figure D.1: Schematic of the cooling plate model considered, showing the primary components referenced: top plate (1), fluid channel (2), and bottom plate (3).Above is the eight-cell module with a typical temperature contour and an arrow showing the coolant flow direction.

Dimensionless data points are defined by dividing each data set by the maximum value in the set, allowing separate quantities to be plotted on the same axis. Polynomial best fit surfaces are generated for the temperature data to eliminate peaks caused by the residuals from the applied convergence criteria in the numerical calculations. The best fit is achieved with a second order polynomial for the flow rate and a first order polynomial for the inlet temperature. The residuals from the best fit plots are in general highest at low volumetric flow rates, but R-Squared values of greater 0.97 are achieved.

The 3D surface plots combined with the-2D contour plots in Figure 8 exhibit the effects of fluid flow rate and inlet temperature on both $\Delta(\Delta T_{cells})$ and T_{max}, showing the interplay between the quantities of interest.

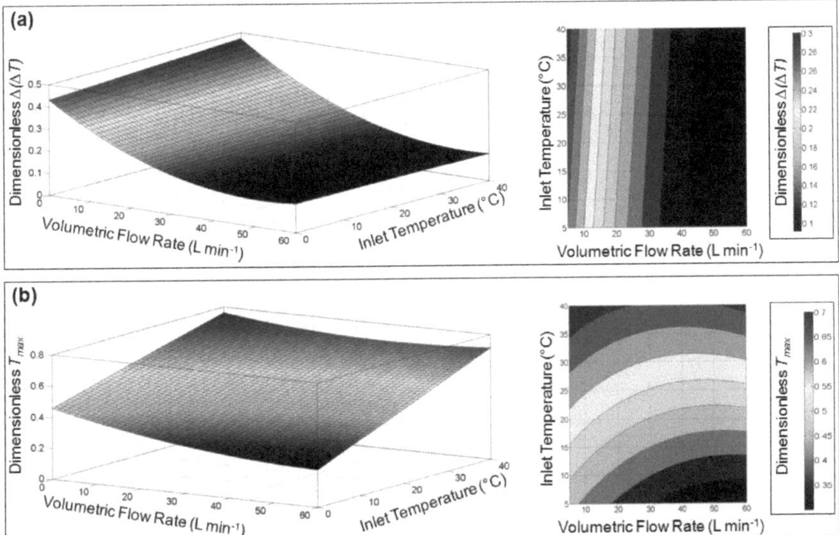

Figure D.2: Comparison of (a) temperature difference $\Delta(\Delta T_{cells})$ and (b) maximum temperature T_{max} in module as a function of fluid flow rate and inlet temperature for a cell heat loss value of 15 W per Cell, each shown on surface and contour plots.

Clearly, optimizing only for $\Delta(\Delta T_{cells})$, T_{max}, or Δp can cause undesirable and even hazardous operating conditions for the battery cells, as flow rate has a greater influence on $\Delta(\Delta T_{cells})$, while inlet temperature has the greatest influence on T_{max}.

The influence of fluid flow rate and inlet temperature on $\Delta(\Delta T_{cells})$ and T_{max} is more evident at higher cell heat loss values, where as the pressure difference (Δp) barely changes as a function of the cell heat loss. The increasing influence of cell heat loss and the corresponding intersections of the $\Delta(\Delta T_{cells})$ and T_{max} surfaces with the Δp best fit surface are shown in Figure 9.

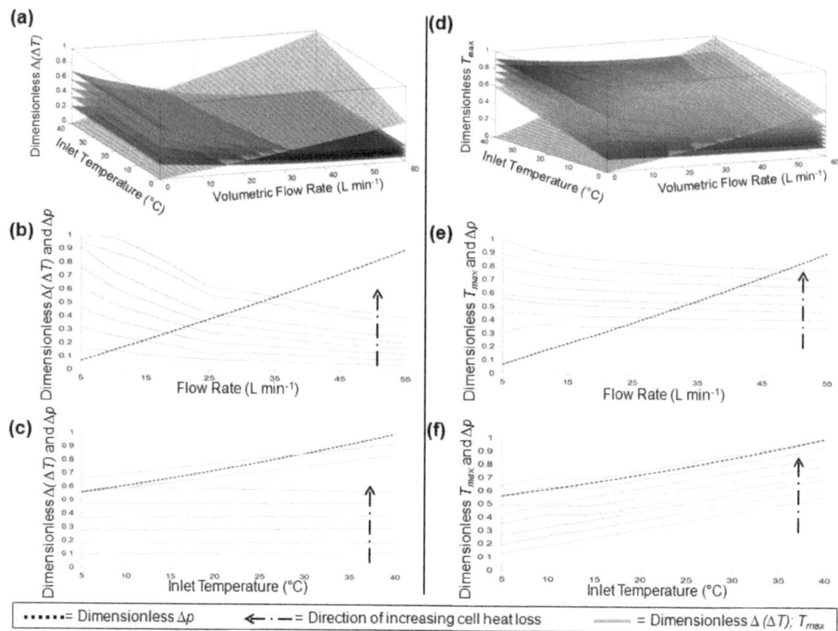

Figure D.3: Intersection of dimensionless best-fit plots of temperature difference $\Delta(\Delta T_{cells})$ with Δp (a) and maximum temperature T_{max} with Δp (d) in a module as a function of fluid flow rate and inlet temperature for various values of cell heat loss from 5 to 40 W per Cell. Additionally shown are best-fit plots of temperature difference $\Delta(\Delta T_{cells})$ with Δp as a function of flow rate (b) and inlet temperature (c) and best-fit plots of maximum temperature T_{max} with Δp as a function of flow rate (e) and inlet temperature (f).

Ultimately, the competing effect of coolant inlet temperature and flow rate on the maximum cell temperature, difference in the individual cell temperature gradients and the coolant pressure drop along the plate can be visualized using the presented analysis method, thereby allowing the thermal system designer to maximize the lifetime and capacity of the battery cells, and ultimately increase vehicle range.

E. Numerical Analysis of the Ceramic Spacers

The results of the numerical simulations pertaining to the ceramic spacer module were shown in [Smith 2015], specific simulations are shown here to help visualize the results obtained in Section 6.

The cooled spacer examined in the feasibility study is rudimentary, assembled from pre-existing components. However, since the concept has proven promising, an optimization is undertaken. To reduce time and cost, computational fluid dynamics (CFD) and a simplified 3D-model are implemented. The model is built to minimize computational time by utilizing symmetry. Spacer size, channel shape, material, thickness, and layout are varied in order to determine the best spacer design. The calculations are performed at various ambient temperatures, coolant temperatures, and coolant flow rates. Cell temperature, standard deviation of the temperature, and the pressure drop over the channel are compared for various spacer designs.

The calculations indicate that the cell temperature is far less sensitive to spacer design variations than pressure drop. This phenomenon exists across various boundary conditions (inlet temperature, flow speed, and ambient temperature).

Interesting trends observed include:

- Channel shape has little influence on mean cell temperature
- Flow direction influences standard deviation of temperature
- The increase in mean cell temperature as more cells are placed between cooled spacers is not linear (i.e. four cells between each spacer versus two).

The large differences in pressure drop shown in Figure E.1 provide an opportunity to reduce the required pumping power while achieving similar thermal performance.

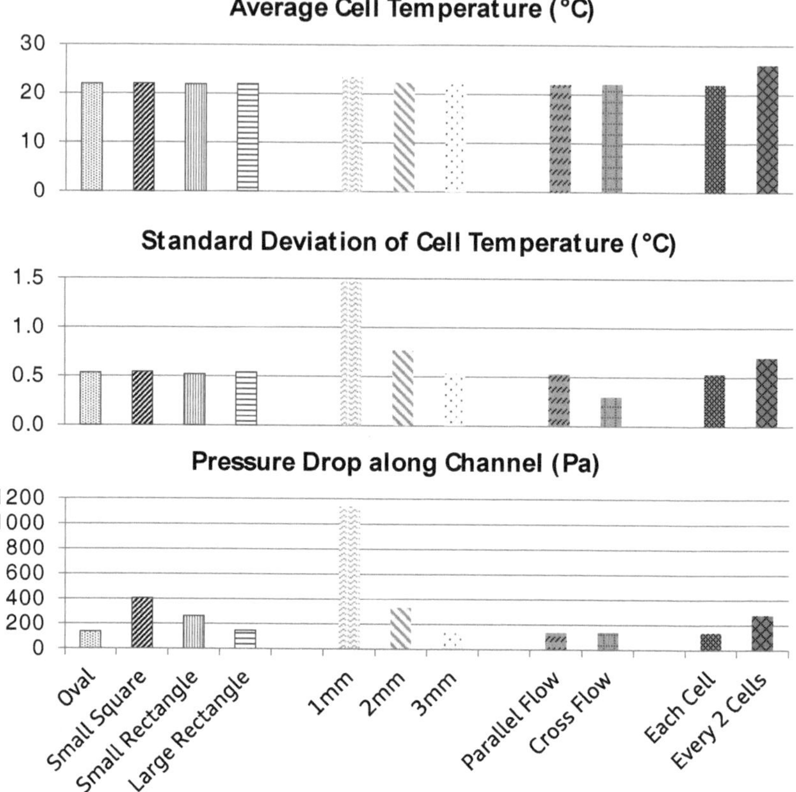

Figure E.1: Comparison of the influence of spacer characteristics and layout on average cell temperature, temperature differences over the cell, and pressure drop along the channel.

Not presented in the work were additional simulations calculated by S. Schüren, which display the inhomogeneous flow pattern causing the temperature differences. In Figure E.2, the wall temperatures of the channels within the spacers are shown, while in Figure E.3, the temperature contours within a cross section of the cells are shown.

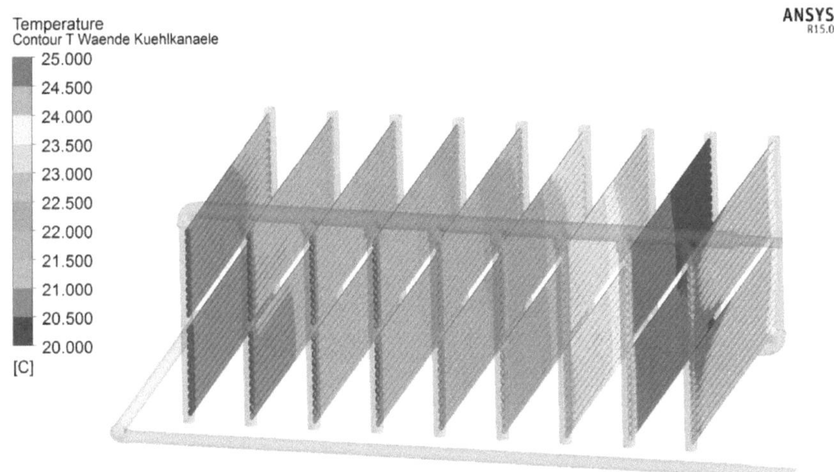

Figure E.2: Wall temperatures of the coolant channels within the nine spacers. The temperature distribution is caused by the flow pattern and distribution amongst the spacers.

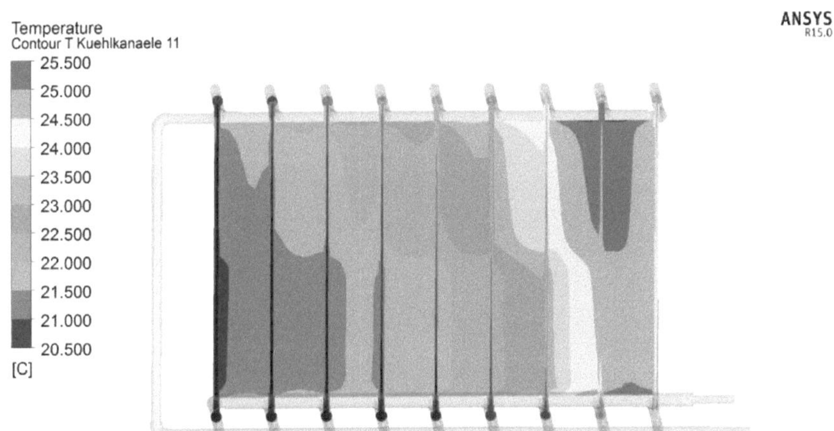

Figure E.3: Temperature contours of the eight cells at the mid plane of the module, as viewed from above. The temperature distribution is caused by the flow pattern and distribution amongst the spacers.